SpringerBriefs in Electrical and Computer Engineering

More information about this series at http://www.springer.com/series/10059

Liang Xiao

Anti-Jamming Transmissions in Cognitive Radio Networks

Springer

Liang Xiao
Xiamen University
Xiamen
Fujian, China

ISSN 2191-8112 ISSN 2191-8120 (electronic)
SpringerBriefs in Electrical and Computer Engineering
ISBN 978-3-319-24290-3 ISBN 978-3-319-24292-7 (eBook)
DOI 10.1007/978-3-319-24292-7

Library of Congress Control Number: 2015954577

Springer Cham Heidelberg New York Dordrecht London

Printed on acid-free paper

Springer International Publishing AG Switzerland is part of Springer Science+Business Media (www.springer.com)

Preface

This book is focused on anti-jamming transmissions in cognitive radio networks (CRNs) and covers several recent research hot topics in this field. First, we present the transmissions based on uncoordinated spread spectrum to address smart jammers in CRNs. We also apply game theory to investigate the interactions between secondary users and jammers and provide game theoretic solutions to suppress jamming incentives in CRNs. Professionals and researchers working in networks, wireless communications, and information technology will find *Anti-jamming Transmissions in Cognitive Radio Networks* valuable material as a reference guide. Advanced-level students studying electrical engineering and computer science will also find this brief a useful study guide.

In Chap. 1, we briefly introduce jamming attacks in cognitive radio networks, focusing on smart jammers and the applications in broadcast services.

In Chap. 2, we review spread spectrum techniques, which have been used by radio communications to counteract jamming attacks for decades and investigate the new challenges that they have to address to protect cognitive radio networks.

In Chap. 3, we present uncoordinated spread spectrum techniques in the anti-jamming communications and investigate the anti-jamming broadcast in CRNs based on node collaboration and uncoordinated spread spectrum. The spectrum efficiency of this technique is analyzed and related implementation issues are reviewed.

In Chap. 4, we analyze the interactions between jammers and secondary users based on game theory. We first formulate the jamming process in CRNs as static games and present their Nash equilibrium and Stackelberg equilibrium. Next, we consider the impact of relay nodes in the anti-jamming transmissions in CRNs and formulate the repeated interactions as a dynamic jamming game, in which the evolutionary equilibrium is analyzed. Finally, we investigate the impact of the inaccurate and incomplete signals on the jamming games.

In Chap. 5, we review important game theoretic mechanisms to improve the performance of the equilibrium in the games, including price mechanisms, auctions, reputations, and trust. Both direct reciprocity and indirect reciprocity principles are illustrated.

In Chap. 6, we present several game theoretic anti-jamming solutions for CRNs, exploiting the requirement for network by secondary users to suppress the jamming motivation of insider attackers and reduce the jammer population in large-scale cognitive radio networks. In addition, reinforcement learning techniques such as Q-learning and WolF are applied to improve the performance of cognitive radio networks against jamming.

In Chap. 7, we conclude this book with a summary and point out several promising research topics in anti-jamming communications of CRNs.

This book could not have been made possible without the contributions by the following people: K.J. Ray Liu, Yan Chen, Weihua Zhuang, Huaiyu Dai, Peng Ning, Vincent Poor, Narayan Mandayam, Chengzhi Li, Wanyi S. Lin, Jinliang Liu, Yan Li, Guolong Liu, Tianhua Chen, Qiangda Li, Changhua Zhou, Yanda Li, Guiquan Chen, Shan Kang, Yuliang Tang, and Lianfen Huang. We would also like to thank all the colleagues whose work enlightened our thoughts and research made this book possible.

Xiamen, Fujian, China   Liang Xiao

Contents

Chapter 1
Introduction

1.1 Cognitive Radio Networks

Cognitive radio (CR) techniques have been developed to address the spectrum shortages of wireless networks and become important for future wireless communications. In cognitive radio networks (CRNs), secondary users (SUs) apply cognitive radio techniques such as spectrum sensing to choose their transmission strategies, such as their transmit power and channel, to avoid interfering with the transmissions of primary users (PUs) that usually ignore the existence of SUs. In this book, we consider a CRN consisting of N SUs and several PUs in the presence of jammers at various locations, as shown in Fig. 1.1.

1.2 Jamming Attacks

Because of the broadcast nature of radio propagation, wireless networks are highly vulnerable to jamming attacks, where jammers aim at interrupting the ongoing legitimate information exchange by injecting replayed or faked signals into wireless media [1]. Consequences of jamming attacks include the degradation of network throughputs, power waste of radio nodes and even denial of service (DoS) attacks in wireless networks.

Jamming attacks can be easily launched in wireless communications, and cannot be addressed by conventional cryptography. Spread spectrum techniques, such as direct sequence spread spectrum (DSSS) and frequency hopping (FH), have been commonly used for decades to counteract jamming [2]. The key idea behind the traditional spread spectrum based anti-jamming techniques is that the senders and

© The Author(s) 2015   1
L. Xiao, *Anti-Jamming Transmissions in Cognitive Radio Networks*,
SpringerBriefs in Electrical and Computer Engineering,
DOI 10.1007/978-3-319-24292-7_1

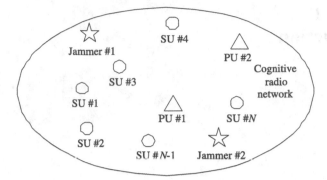

Fig. 1.1 Anti-jamming communications in cognitive radio networks, consisting of N secondary users, two primary users and two jammers

legitimate receivers share the same spreading codes in DSSS or frequency hopping patterns in FH, which can be viewed as physical-layer pre-shared secrete keys that are unknown to jammers.

In this book, omniscient jammers with bounded computation and transmission capability are considered, each transmitting on one or several channels for a long enough time in a time slot to effectively block a packet. Jamming attacks are usually categorized into non-responsive and responsive ones, based on the attempts of a jammer to detect the ongoing transmission before sending jamming signals.

Typical non-responsive jamming strategies in wireless systems with multiple channels include constant jamming, random jamming and sweep jamming. Constant jammers block the same channels all the time of interests. Random jammers switch their jamming channels randomly. Sweep jammers sweep the whole channel range of the wireless system over a long time.

In contrast, responsive jammers sense the channels before sending jamming signals, as it usually takes less time to switch the sensing channels than to change the jamming channels. As a powerful type of jammers, a responsive-sweep jammer conducts both non-responsive and responsive jamming independently and simultaneously, whose jamming strength depends on both the total number of jammed channels and the number of sensed channels in a time slot [3, 4]. The analysis based on the responsive-sweep jamming provides a lower bound for the performance of cognitive radio networks.

1.2.1   Smart Jammers

As a special type of insider attacks, smart jammers have more flexible jamming strategies and stronger jamming strengths compared with traditional jammers. As shown in Fig. 1.2, a smart jammer first observes the ambient radio environment before actually sending jamming signals on the chosen channels with selected

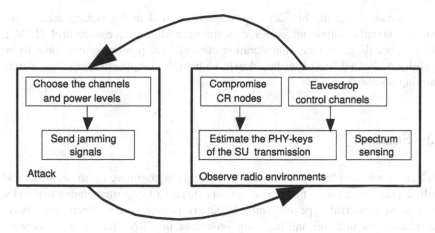

Fig. 1.2 Flowchart of smart jamming in cognitive radio networks, consisting of two stages: attack and observation

jamming power levels. By compromising some CR nodes or eavesdropping the public control channels of the CRN, a smart jammer can estimate the PHY-layer keys of the spread spectrum systems, such as the frequency hopping pattern of the CR transmitter and thus efficiently block its following transmissions. Similar to the other types of responsive jammers, smart jammers can also adjust their working channels based on the spectrum sensing results. In addition, smart jammers in a large-scale CRN can also apply other strategies, such as to open fake relay channels to lure the receivers and thus block them more efficiently [5].

1.2.2 Jamming in Broadcast

Jamming-resistant broadcasts are not only important for safety-critical applications such as emergency alert broadcast, but also critical for the distribution of important network information such as public keys and control information in the networks. One key vulnerability of the conventional anti-jamming techniques is the requirement of pre-shared secrete keys, such as spreading codes or frequency hopping pattern at the senders and receivers. This requirement suffers from scalability concerns of broadcast, and may even be unfeasible in dynamic networks with compromised receivers. This problem was recognized recently, leading to a series of promising research efforts, such as uncoordinated FH [3, 6] techniques, uncoordinated DSSS [7, 8], and BBC [9].

In the broadcast in cognitive radio networks, a source node aims to convey a message consisting of several packets to all the nodes in the network, possibly by multiple hops. Compared with pairwise communications, broadcast incurs more challenges as well as chances especially in the large-scale networks. For example, receivers can obtain the broadcast message if located in the outage area of the jamming signals. Important issues concerning multiple access control (MAC), packet scheduling, transmission duration control, and power control have to be carefully addressed to enable broadcasts with high energy efficiency and strong jamming resistance.

1.3 Summary

In Chap. 1, we have briefly reviewed the concepts in cognitive radio networks and studied jamming attacks that throw serious threats to cognitive radio networks. As a most powerful type of jammers, smart jammers have been introduced. The challenges that the anti-jamming broadcast in CRNs has to address were investigated and related work was reviewed.

References

1. Poisel, R.A.: Modern Communications Jamming Principles and Techniques. Artech House, Boston (2006)
2. Goldsmith, A.: Wireless Communications. Cambridge University Press, Cambridge (2005)
3. Strasser, M., Popper, C., Capkun, S., Cagalj, M.: Jamming-resistant key establishment using uncoordinated frequency hopping. In: Proceedings of IEEE Symposium on Security and Privacy, pp. 64–78 (2008)
4. Strasser, M., Popper, C., Capkun, S.: Efficient uncoordinated FHSS anti-jamming communication. In: Proceedings of ACM International Symposium Mobile Ad Hoc Networking and Computing (MobiHoc), pp. 207–218 (2009)
5. Xiao, L., Dai, H., Ning, P.: Jamming-resistant collaborative broadcast using uncoordinated frequency hopping. IEEE Trans. Inf. Forensics Secur. 7(1), 297–309 (2012)
6. Slater, D., Tague, P., Poovendran, R., Matt, B.: A coding-theoretic approach for efficient message verification over insecure channels. In: Proceedings of ACM Conference of Wireless Network Security (WiSec), pp. 151–160 (2009)
7. Popper, C., Strasser, M., Capkun, S.: Jamming-resistant broadcast communication without shared keys. In: Proceedings of USENIX Security Symposium, pp. 231–248 (2009)
8. Liu, Y., Ning, P., Dai, H., Liu, A.: Randomized differential DSSS: jamming-resistant wireless broadcast communication. In: Proceedings of IEEE International Conference on Computer Communication (INFOCOM), pp. 1–9 (2010)
9. Baird, L., Bahn, W., Collins, M., Carlisle, M., Butler, S.: Keyless jam resistance. In: Proceedings of IEEE Information Assurance and Security Workshop, pp. 143–150 (2007)

Chapter 2
Spread Spectrum-Based Anti-jamming Techniques

2.1 Introduction

Cognitive radio techniques have been developed to provide dynamic spectrum access and improve spectral efficiency. The broadcast nature of radio propagation and random spectrum access of SUs, however, result in high vulnerability of CRNs to jamming attacks [1]. In CRNs, secondary users are authorized to access licensed channels without interfering with primary users. Jamming attacks can lead to denial-of-service attacks that deny users from spectrum access [2]. Anti-jamming communication is important for many safety-critical applications such as emergency alert transmission and navigation signal dissemination, and is critical for the distribution of important information such as the public key and system control information in wireless systems.

Jamming attacks can be easily launched in cognitive radio communications. Spread spectrum techniques, including direct sequence spread spectrum and frequency hopping, have been widely used for decades to counteract jamming. These techniques require pre-shared secret keys (such as spreading codes in DSSS or frequency hopping patterns in FH) at the senders and legitimate receivers. These secret keys enable the sender to spread the signal such that its transmission becomes unpredictable for the jammer, thus reducing the probability of jamming attacks. Therefore, this chapter aims to provide insights on the anti-jamming technologies based on DSSS and FH, and we also discuss potential challenges of jammers in CRNs.

© The Author(s) 2015   5
L. Xiao, *Anti-Jamming Transmissions in Cognitive Radio Networks*,
SpringerBriefs in Electrical and Computer Engineering,
DOI 10.1007/978-3-319-24292-7_2

2.2   Frequency Hopping

In frequency hopping systems, a transmitter sends signals over different channels (carrier frequencies) from a given channel set according to a hopping pattern which is known by both the transmitter and the receiver in advance. Therefore, the signal is transmitted on one frequency for a certain period of time, and then sent on another frequency. The receiver has to use the same center frequency and keep synchronization with the transmitter. As shown in Fig. 2.1, the frequency hopping system sends packets over one of N channels.

There are two types of frequency hopping techniques: slow frequency hopping and fast frequency hopping. In slow frequency hopping systems, the transmitted signal changes channels for one or more data bits. In contrast, each data bit is distributed over multiple channels in fast frequency systems. Slow frequency hopping has been used in IEEE 802.11b standards [3]. Fast frequency hopping can be used in multiple frequency shift keying modulation [4]. Frequency hopping helps protect transmissions from fading and interference [5]. Frequency hopping based on parallel spectrum sensing without interruption was proposed to address the conflict between the quality of service satisfaction and reliable spectrum sensing for licensed users in IEEE 802.22 [6].

Uncoordinated frequency hopping (UFH) was proposed to break the dependency on pre-shared key in [7], in which the transmitter and receiver select the channels randomly and independently from a given channel set. In UFH, each packet is sent over a channel randomly selected. BMA scheme incorporates one-way authentication and error control coding to improve the efficiency of UFH [8]. USD-FH system was proposed in [9], in which a Diffie-Hellman key establishment

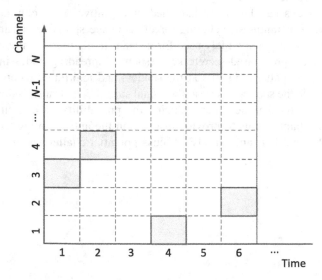

Fig. 2.1 Illustration of a frequency hopping system over N channels

message is transmitted with uncoordinated frequency hopping, and the data message is transmitted using coordinated FH with the established FH pattern. Delay-bounded adaptive UFH proposed in [10] addresses adaptive jamming and applies online optimization to improve the performance of UFH. Online adaptive UFH was formulated as a multi-armed bandit problem in [11]. In collaborative UFH-based broadcast system, the nodes that have received the broadcast message relay the message to improve the communication efficiency and jamming resistance [12].

2.3 Direct Sequence Spread Spectrum

As a widely used spread spectrum technique, DSSS applies spreading codes such as pseudo noise (PN) code to modulate data sequences, and is also known as direct sequence code division multiple access (DS-CDMA) [5]. In DSSS systems, the data signal is multiplied by the spreading code that consists of pulses with shorter durations over a wider bandwidth based on the spreading factor. Upon receiving the spreading code, the receiver uses the same spreading sequence over the received signal to reconstruct the original data [13].

The transmitted signal as a noise signal obtains stronger interference resistance, and can recover from the damages in transmissions. Different spreading sequences permit multiple users to share the same channel. By spreading signals with spreading codes, DSSS system can reduce the influence of background noise on transmissions. Compared to frequency hopping techniques, DSSS reduces the interference all the time [5]. DSSS systems usually achieve more reliable communication performance, while devices in frequency hopping systems are cheaper and consume less power in general.

DSSS has been widely used in wireless communication systems such as IEEE 802.11b [14], satellite navigation systems [15], Galileo [16], GLONASS [17], and ultra-wideband systems [18]. Uncoordinated DSSS was proposed to avoid the dependence on pre-shared keys and improve the jamming resistance [19].

2.4 Challenges of Jammers in CRNs

Due to the opportunistic access and broadcast nature of cognitive radio networks, secondary users have to address jammers, especially advanced smart jammers in dynamic and distributed networks without interfering with primary users. Anti-jamming techniques based on frequency hopping have been investigated in [11] and game theory has been used to formulate anti-jamming transmissions in [20]. Jamming attacks have been analyzed for decades in wireless systems [10, 21]. The anti-jamming game was formulated to investigate the jamming strategies and defense policies in [22].

The traditional anti-jamming transmissions based on spread spectrum assume that both the transmitter and the receiver share the spreading codes or frequency hopping patterns in advance, which have to be hidden from jammers. Unfortunately, smart jammers can utilize this dependency to increase their jamming strengths. As an advanced type of reactive jammers that can sense spectrum before jamming, smart jammers can not only block the target channels with flexible power, but also eavesdrop the public control channels and compromise CR nodes in large-scale CRNs to derive the spreading codes of the transmitter.

Uncoordinated frequency hopping [7] and its variations [9, 11, 19] were proposed to address smart jammers. Without requiring any pre-shared FH pattern, a receiver in UFH randomly selects a channel from the public channel set. A packet can be successfully received if both the transmitter and the receiver come across the same unblocked channel, indicating a very small packet reception rate due to the large number of channels of UFH to counteract jamming. Collaborative UFH-based broadcast exploits node cooperations to provide the spatial and spectrum diversity in single-hop networks [11] and multi-hop networks [12], and thus enhances the communication efficiency of UFH [23]. The efficient communication without pre-shared key proposed in [24] uses intractable forward-decoding and efficient backward-decoding to decrease the energy cost compared with UFH. In spite of all these efforts, the anti-jamming transmissions based on UFH still suffer from low communication efficiency, which has to be further studied in the future.

2.5 Summary

Jamming attacks can lead to denial-of-service attacks in cognitive radio networks. Spread spectrum techniques such as frequency hopping and DSSS have been widely used to address jamming. However, due to the requirement of pre-shared PHY-layer secret keys between the transmitter and the receiver, the anti-jamming broadcast in large-scale dynamic CRNs is still vulnerable to smart jamming.

References

1. Poisel, R.A.: Modern Communications Jamming Principles and Techniques. Artech House, Boston (2006)
2. Attar, A., Tang, H., Vasilakos, A.V., Yu, F.R., Leung, V.: A survey of security challenges in cognitive radio networks: solutions and future research directions. Proc. IEEE 100(12), 3172–3186 (2012)
3. Yang, L.-L., Hanzo, L.: Slow frequency-hopping multicarrier DS-CDMA for transmission over Nakagami multipath fading channels. IEEE J. Sel. Areas Commun. 19(7), 1211–1221 (2001)
4. Fiebig, U.C.G.: Iterative interference cancellation for FFH/MFSK MA systems. Proc. IEE Commun. 143, 380–388 (1996)
5. Goldsmith, A.: Wireless Communications. Cambridge University Press, Cambridge (2005)

6. Hu, W., Willkomm, D., Abusubaih, M., et al.: Cognitive radios for dynamic spectrum access-dynamic frequency hopping communities for efficient IEEE 802.22 operation. IEEE Commun. Mag. **45**(5), 80–87 (2007)
7. Strasser, M., Capkun, S., Cagalj, M.: Jamming-resistant key establishment using uncoordinated frequency hopping. In: Proceedings of IEEE Symposium on Security and Privacy, pp. 64–78 (2008)
8. Strasser, M., Pöpper, C., Čapkun, S.: Efficient uncoordinated FHSS anti-jamming communication. In: Proceedings of ACM International Symposium on Mobile Ad Hoc Networking and Computing, pp. 207–218 (2009)
9. Liu, A., Ning, P., Dai, H., Liu, Y., Wang, C.: USD-FH: Jamming-resistant wireless communication using frequency hopping with uncoordinated seed disclosure. In: Proceedings of IEEE International Conference on Mobile Adhoc and Sensor Systems (MASS), pp. 41–50 (2010)
10. Wang, Q., Xu, P., Ren, K., Li, M.: Delay-bounded adaptive UFH-based anti-jamming wireless communication. In: Proceedings of IEEE International Conference on Computer Communication (INFOCOM), pp. 1413–1421 (2011)
11. Wang, Q., Xu, P., Ren, K., Li, X.-Y.: Towards optimal adaptive UFH-based anti-jamming wireless communication. IEEE J. Sel. Areas Commun. **30**(1), 16–30 (2012)
12. Xiao, L., Dai, H., Ning, P.: Jamming-resistant collaborative broadcast using uncoordinated frequency hopping. IEEE Trans. Inf. Forensics Secur. **7**(1), 297–309 (2012)
13. Rappaport, T.S.: Wireless Communications: Principles and Practice, vol. 2. Prentice Hall, Upper Saddle River (1996)
14. Seide, R.: Capacity, coverage, and deployment considerations for IEEE 802.11g. Cisco systems white paper, San Jose (2003)
15. Kaplan, E., Hegarty, C.: Understanding GPS: Principles and Applications. Artech House, Boston (2005)
16. Hein, G.W., Godet, J., Issler, J.-L., Martin, J.-C., Lucas, R., Pratt, T.: The Galileo frequency structure and signal design. In: Proceedings of Institute of Navigation GPS, pp. 1273–1282 (2001)
17. Ellingson, S.W., Bunton, J.D., Bell, J.F.: Removal of the GLONASS C/A signal from OH spectral line observations using a parametric modeling technique. Astrophys. J. Suppl. Ser. **135**(1), 87–93 (2001)
18. Foerster, J.: The performance of a direct-sequence spread ultrawideband system in the presence of multipath, narrowband interference, and multiuser interference. In: Proceedings of IEEE Conference on Ultra Wideband Systems and Technologies, pp. 87–91 (2002)
19. Popper, C., Strasser, M., Capkun, S.: Anti-jamming broadcast communication using uncoordinated spread spectrum techniques. IEEE J. Sel. Areas Commun. **28**(5), 703–715 (2010)
20. Xiao, L., Dai, H., Ning, P.: Jamming-resistant collaborative broadcast in wireless networks, part II: multihop networks. In: Proceedings of IEEE Global Communications Conference (GLOBECOM), pp. 1–6 (2011)
21. Xu, W., Trappe, W., Zhang, Y., Wood, T.: The feasibility of launching and detecting jamming attacks in wireless networks. In: Proceedings of ACM International Symposium on Mobile Ad Hoc Networking and Computing, pp. 46–57 (2005)
22. Wu, Y., Wang, B., Liu, K.R., Clancy, T.C.: Anti-jamming games in multi-channel cognitive radio networks. IEEE J. Sel. Areas Commun. **30**(1), 4–15 (2012)
23. Li, C., Dai, H., Xiao, L., Ning, P.: Analysis and optimization on jamming-resistant collaborative broadcast in large-scale networks. In: Proceedings of IEEE International Conference on Signals, Systems and Computers (ASILOMAR), pp. 1859–1863 (2010)
24. Cassola, A., Jin, T., Noubir, G., Thapa, B.: Efficient spread spectrum communication without preshared secrets. IEEE Trans. Mob. Comput. **12**(8), 1669–1680 (2013)

Chapter 3
Anti-jamming Techniques Based on Uncoordinated Spread Spectrum

3.1 Introduction

Traditional anti-jamming communications mostly rely on spread spectrum techniques, including DSSS and FH, in which pre-shared PHY-layer secret keys, such as spreading code sequences and frequency hopping patterns, are assumed to be known by the transmitter and receiver. The requirement on the pre-shared key makes it challenging to distribute the PHY-layer keys to all the receivers in a large scale mobile network with nodes entering and leaving. In addition, compromised CR nodes can eavesdrop the public control channels to obtain the broadcast spread spectrum patterns and thus efficiently block the following transmissions.

Uncoordinated spread spectrum techniques enable anti-jamming transmissions without requiring any pre-shared keys, but suffer from a low communication efficiency due to the lack of coordination between the sender and receiver. For instance, the relative throughput of the UFH compared with coordinated FH is only on the order of 10^{-3} for a spreading ratio of 200 [3]. Therefore, the erasure coding and one-way authentication based on bilinear maps were applied to improve the communication efficiency of UFH [1]. In addition, USD-FH scheme was proposed in [4] to further improve the efficiency and robustness of UFH, in which the hopping pattern is conveyed through UFH to allow the message transmissions via coordinated FH. Collaborative UFH (CUFH) was proposed in [5] to increase the throughput and enhance jamming resistance, in which nodes having obtained the message serve as relays for the remaining nodes to expedite the broadcast process against smart jammers.

© The Author(s) 2015
L. Xiao, *Anti-Jamming Transmissions in Cognitive Radio Networks*,
SpringerBriefs in Electrical and Computer Engineering,
DOI 10.1007/978-3-319-24292-7_3

3.2 Uncoordinated Spread Spectrum

We investigate the uncoordinated spread spectrum techniques to address jamming attacks in cognitive radio networks. Based on the randomly selected PHY-layer keys, neither receivers nor jammers know the keys in advance. Thus jammers with limited jamming energies cannot efficiently block the transmissions of secondary users. In particular, each transmitter in UDSSS randomly selects a spreading code from a set of code sequences, while each receiver decodes the packets through brute-force search. In UFH, each transmitter (or receiver) randomly chooses a channel from the public known channel pool to transmit (or receive) a packet. The receiver obtains a packet only if the transmitter and the receiver happen to use the same channel. On the other hand, the jamming strength against UDSSS systems mostly depends on the computational speed of the jammers, while the performance of a UFH system mostly relies on the capability of the jammers to sense and switch their frequency channels.

3.2.1 Uncoordinated Frequency Hopping

In uncoordinated frequency hopping systems, a message is divided into several short packets, each transmitted over a selected channel only known to the sender. The sender and the receiver randomly and independently choose a channel from the public channel set. Successful transmission is possible if both nodes come across on the same channel. The packets are sent sequentially and repeatedly for a sufficiently long time. UFH scheme is resistant to packet loss, and can establish an unanticipated and spontaneous communication without pre-shared keys. As illustrated in Fig. 3.1, Alice aims to send Bob several fragment messages. Both the secondary users randomly choose a channel from the public C channels in each time slot. To relax the synchronization requirement, Bob changes its channels at a slower rate than Alice. The several packets are sent sequentially and repeatedly, until Bob successfully receives them and reassembles these packets with a collision-resistant hash function [3].

3.2.2 Uncoordinated DSSS

In UDSSS, the transmitter randomly selects a spreading code sequence from a public and known set of spreading sequences for each message. The receiver uses the trial-and-error method to despread the received messages by randomly applying a spreading code sequence. A message is accepted, if the receiver chooses the same spreading code and successfully verifies its signature. For example, in Fig. 3.2, the sender aims to transmit l fragment messages, denoted by M_i with $1 \leq i \leq l$, based

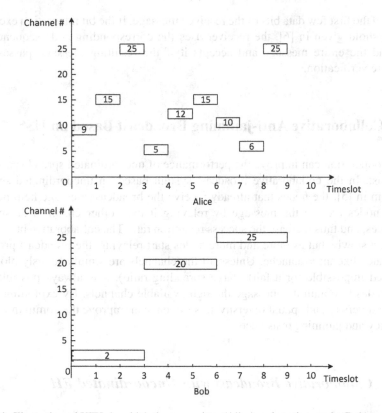

Fig. 3.1 Illustration of UFH, in which the transmitter (Alice) and receiver node (Bob) randomly and independently choose a frequency channel from C channels in each time slot

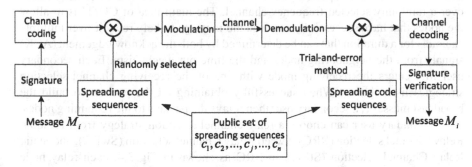

Fig. 3.2 Transmission block of uncoordinated DSSS

on the public code set including n orthogonal spreading code sequences, denoted by c_j, with $1 \leq j \leq n$. The sender first derives the digital signature for the message using its private key K_S and then randomly selects a spreading sequence c_j. Each message has to be repeatedly transmitted, each with a new spreading code sequence randomly selected. Upon receiving a signal, the receiver chooses a code sequence to despread the first few data bits of the received message. If the bit integration exceeds the threshold given in [6], the receiver uses the corresponding code sequence to despread the entire message and accepts it if the resulting message passes the signature verification.

3.3 Collaborative Anti-jamming Broadcast Based on USS

Node cooperation can improve the performance of uncoordinated spread spectrum broadcast. In the collaborative broadcast system based on uncoordinated spread spectrum in [5], the nodes that already receive the broadcast message help neighboring nodes receive the message by relaying it over other channels or spread sequences, and thus increase the success reception rate. The collaborative broadcast may start slowly, but as more and more nodes start relaying, the broadcast process accelerates like an avalanche. Unless all the channels are simultaneously blocked (assumed impossible for a fairly large spreading ratio), it is always possible for some nodes to obtain the message through available channels. By exploiting both spectral diversity and spatial diversity, this system can improve the communication efficiency and jamming resistance.

3.3.1 Collaborative Broadcast with Uncoordinated FH

In the collaborative UFH-based broadcast (CUFH) as shown in Fig. 3.3, the source node sequentially and repeatedly sends the packets of the broadcast message, each over a randomly selected frequency channel. The main idea of CUFH is to allow the nodes that have successfully received the message to help relay it over multiple channels for a duration that can be determined by both the acknowledgement (ACK) signals from the neighboring nodes and the time-out mechanism. Each secondary user first enters the receiving mode with one of the receiving channel selection strategies listed below. When successfully obtaining all the packets to build the broadcast message, a secondary user then relays the packets to the remaining nodes.

A secondary user can choose a relay channel selection strategy from Random Relay Channel selection (RRC), Sweep Relay Channel selection (SwRC), and Static Relay Channel selection (StRC). In RRC as shown in Fig. 3.4, each relay node randomly and independently selects one out of the C channels to send a packet, similar to the source node. This strategy is amenable to distributed implementation and has good scalability, while sometimes some relay nodes, as well as the source node, may happen to select the same channel, leading to collision and failure of transmission. Even with perfect synchronization and collaboration among the source and relays such that the same packet is broadcast by all relays and the source at the same time, such overlap still leads to the waste of energy and reduced opportunity.

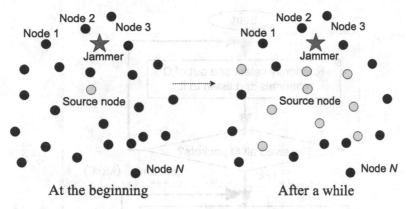

Fig. 3.3 Illustration of collaborative anti-jamming broadcast, in which the nodes that already receive the message help relay the information

To evaluate the theoretical maximum of RRC, we consider the idealized version SwRC. The relays with SwRC take non-overlapping channels for each packet transmission: The first relay node randomly selects one from C channels, the second relay node randomly chooses one out of the remaining $C - 1$ channels, and so on. This approach was proposed mainly as an alternative for RRC to facilitate the discussion on the trade-off between performance and complexity (the protocol overhead for coordination). The SwRC strategy avoids the possible collision incurred in RRC, but requires information exchange among local nodes to determine the non-overlapping channel in the broadcast.

In contrast to RRC and SwRC where the relay channels change randomly from one time slot to another, the relays with the StRC strategy take fixed non-overlapping channels through the message broadcast process as shown in Fig. 3.5. The nodes are assumed to have been preassigned unique IDs, which, together with a suitable algorithm (see, e.g., [7]), guarantees negligible channel collisions. Each node is assumed to know the IDs of the neighboring nodes within its communication distance, and hence the IDs of all potential relay channels in its area, which constitute its initial relay channel list. Since a fixed set of relay channels are employed during the whole message broadcast process, it is reasonable to assume that the relay channels are (after some time) known to both the yet-to-inform receivers and jammers.

At a first glance, the StRC strategy seems to be a dumb approach: The jammers can go ahead to block these relay channels even without sensing, and thus the hope of the receivers still lies in the UFH-based source node transmission. Actually this approach captures the essence of collaborative broadcast. As long as all the channels are not blocked simultaneously, the number of relays increases with time. When the turning point is reached so that the jammers can no longer block all relay channels, the communication efficiency will be boosted dramatically.

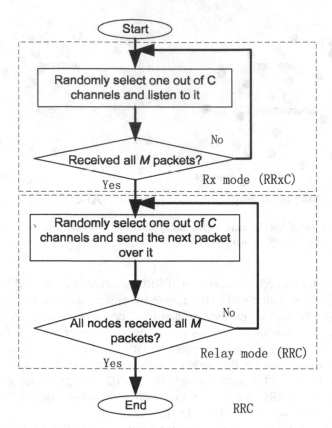

Fig. 3.4 Flowchart of the collaborative anti-jamming broadcast of a cognitive radio node with RRxC and RRC

An alternative view is that, as the UFH-based source node already provides uncertainty in its channel selection to counteract jamming, the StRC strategy introduces certainty in the relay selection to improve the communication efficiency. Furthermore, the StRC strategy is also easy to implement, as it saves the efforts of channel switching, and requires little communication overhead for coordination.

In a Random Receiving Channel selection (RRxC) scheme, each receiver hops randomly and independently over the C channels. For the StRC relay strategy, it also devises an Adaptive Receiving Channel selection (ARxC) strategy. As mentioned, each node is assumed to know the initial relay channel list. After listening to a potential relay channel for a sufficient time, a node can determine whether the channel is clear, active (relaying packets), or jammed. Each receiver with the ARxC strategy first continuously sweeps its relay channel list, one at a time, in an order only known to itself. When encountering an active channel, the node can receive a packet there. If all the relay channels are jammed, the node switches to the RRxC mode. In the RRxC mode, when coming across a clear or active relay channel, the node restricts itself to the relay channels again. This process repeats until a receiver successfully obtains all the packets.

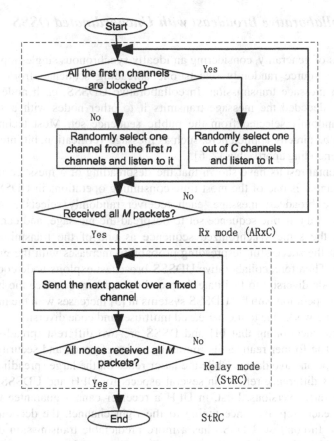

Fig. 3.5 Flowchart of the collaborative anti-jamming broadcast of a cognitive radio node with ARxC and StRC

In essence, a receiver taking the ARxC strategy first attempts to take advantage of the available relay channels. However, these relay channels are also known to the jammers, and it is definitely in the jammers interest to first block them. Under strong jamming such that all known relay channels are blocked, a receiver then switches back to RRxC. However, instead of continuously jamming the static relay channels, the jammers may just spend enough energy to spoof the receivers away, and hence a smart receiver has to check for such scenarios and come back to relay channels.

Packet verification techniques [1, 8] and message authentication techniques on the application layer [2] can be employed to identify such fake channels, and remove them from the active relay list. The relay channel list is continuously updated through the broadcast process, and the dual-mode operation (ARxC and RRxC) achieves a good balance between communication efficiency and security.

3.3.2 Collaborative Broadcast with Uncoordinated DSSS

Without loss of generality, considering an ideally synchronous single-hop wireless network, the source randomly selects one spreading sequence from a public set for each message transmission. In collaborative UDSSS, each node that has successfully decoded the message transmits it to other nodes with a spreading sequence randomly selected from the public sequence set. Most techniques in UDSSS can be directly applied here, such as message verification, bit interleaving, and packet encoding approaches [2, 6].

Experimental results have shown that the despreading of a message in a trial-and-error manner is one of the most time-consuming operations in UDSSS. Once recording the broadcast message, each receiver randomly selects a spreading sequence from the public sequence set to despread the message, and can succeed if choosing the same synchronized sequence as one of the transmitters. It is intuitive that the successful despreading probability increases with the number of transmitters. Therefore, collaborative UDSSS broadcast exploits node cooperation to provide code diversity to facilitate despreading and hence reduce the broadcast delay. The cooperation gain for UDSSS systems in [5] increases with the number of nodes in the network, due to the increased multiuser and code diversity.

It is worth mentioning that FH and DSSS are two different spread-spectrum techniques. The former realizes its immunity to interference and security attacks through escape and avoidance, while the latter relies on the large spreading gain to mitigate. This difference reflects in several aspects for UFH and UDSSS, and for their collaborative versions. First, in UFH a receiver cannot guarantee to obtain a packet in each hop, but once tuning to the right channel, the decoding effort is minimum. In contrast, UDSSS has a more predictable transmission delay but requires significantly more decoding efforts. Therefore, collaborative broadcast help UFH more on the transmission side, and UDSSS more on the reception side. It should be noted that collaborative UDSSS broadcast also introduces a higher interference level to receivers. Due to their inherent differences, the performance of UFH and collaborative UFH broadcast is mainly limited by the hardware capability (such as sensing and switching) and available power of legitimate nodes and jammers, while that of UDSSS and collaborative UDSSS, among others, is mainly restricted by the computing power.

3.4 Spectrum Efficiency

We first analyze the spectrum efficiency of the collaborative UFH-based broadcast in a simplified snapshot scenario against responsive-sweep jamming in a single-hop network. The successful reception rate of each receiver, denoted by $p_a(n)$, is the probability for a receiver to successfully receive a packet in a given time slot, in the network with a source node, a jammer with jamming probability denoted by

p_J, and n relay nodes. All $n + 1$ transmitters are perfectly synchronized so that a receiver can obtain the packet even if it is sent simultaneously by multiple SUs, i.e., multiple transmissions on the same channel do not incur conflicts. The successful packet reception rates for different relay and receiving channel selection strategies are given below.

Lemma 3.1. *The successful packet reception rate of the collaborative UFH-based broadcast with RRC and RRxC strategies under ideal synchronization is given by*

$$p_a^{RRC}(n) = \left(1 - \left(1 - \frac{1}{C} \right)^{n+1} \right) (1 - p_J). \qquad (3.1)$$

Proof. The probability for a source or relay node to transmit over a specific channel is $1/C$. With RRC and RRxC strategies, these nodes randomly and independently choose their transmission channels. Hence the probability that none of them picks the same channel with the receiver is $(1 - (1 - 1/C)^{n+1})$. Assuming perfect relay timing and content synchronization, the receiver can obtain the packet, if working on a channel that is clear from jamming with probability of $(1 - p_J)$, and is selected by at least one of these transmitters (with probability of $(1 - (1 - 1/C)^{n+1})$). Hence the successful packet reception rate is $(1 - (1 - 1/C)^{n+1})(1 - p_J)$.

Lemma 3.2. *The successful packet reception rate of the collaborative UFH-based broadcast with SwRC and RRxC strategies under ideal synchronization is given by*

$$p_a^{SwRC}(n) = \left(1 - \left(1 - \frac{1}{C} \right) \left(1 - \frac{n}{C} \right) \right) (1 - p_J). \qquad (3.2)$$

Proof. See [5].

To maximize the average number of blocked packets, the most powerful jamming strategy is to first block as many relay channels as possible, and then continue to attack the non-relay channels, if at all possible. In the collaborative broadcast with StRC, one copy of the packet is sent by a relay node on each of the n relay channels, while only the source node can access the remaining $C - n$ channels. When knowing that the relay nodes perform the StRC strategy, the jammer first blocks as many relay channels as possible. The sensing capability of the jammer does not help block these static known relay channels, and the jamming probability against the relay channels is solely determined by the maximal transmission power and the blocking capability of the jammer. Therefore, the turning point happens at $n_J C_J$, where n_J is number of jamming cycle in a slot and C_J is number of channels concurrently blocked by a jammer and we have the following:

Lemma 3.3. *The successful packet reception rate of the collaborative UFH-based broadcast with StRC and RRxC strategies under ideal synchronization is given by*

$$p_a^{StRC,RRxC}(n) = \begin{cases} \frac{1}{C}\left(n - n_J C_J + 1 - \frac{n}{C}\right), & n > n_J C_J \\ \frac{1-p_J}{C}, & \text{o.w.} \end{cases} \qquad (3.3)$$

Proof. See [5].

Lemma 3.4. *The successful packet reception rate of the collaborative UFH-based broadcast with StRC and ARxC strategies under ideal synchronization is given by*

$$p_a^{StRC,ARxC}(n) = \begin{cases} 1 - \frac{n_J C_J}{n}, & n > n_J C_J \\ \frac{1-p_J}{C}, & \text{o.w.} \end{cases} \qquad (3.4)$$

Proof. See [5].

As indicated in Eqs. (3.3) and (3.4), ARxC exceeds RRxC, if the number of relays overpowers the (hard) jamming capability, i.e., $n > n_J C_J$, which corresponds to the scenario with significant cooperation gains. As shown in Eq. (3.4), the successful packet reception rate rises from $(1 - p_J)/C$ to $1/(n_J C_J + 1)$, as the number of relay nodes increases from $n_J C_J$ to $n_J C_J + 1$.

Based on the successful packet reception rate $p_a(n)$ with n relay nodes for various strategies, we consider the corresponding cooperation gain for perfect relay synchronization, defined as $G(n) \triangleq p_a(n)/p_a(0)$, where the benchmark performance of the noncooperative UFH-based broadcast, $p_a(0) = p_a(n = 0)$. For RRC and SwRC, by Eqs. (3.3) and (3.4), the cooperation gains for sufficiently large C under perfect synchronization can be approximated by

$$G^{RRC}(n) \approx G^{SwRC}(n) \approx n + 1. \qquad (3.5)$$

For the case of StRC, $p_a(0) = (1 - p_J)C$ and by Eqs. (3.2) and (3.4), the cooperation gain for RRxC and ARxC are given, respectively, by

$$G_{RRxC}^{StRC}(n) = \begin{cases} \frac{1}{1-p_J}\left(n - n_J C_J + 1 - \frac{n}{c}\right), & n > n_J C_J \\ 1, & \text{o.w.} \end{cases} \qquad (3.6)$$

and

$$G_{RRxC}^{StRC}(n) = \begin{cases} \frac{C}{1-p_J}\left(1 - \frac{n_J C_J}{c}\right), & n > n_J C_J \\ 1, & \text{o.w.} \end{cases} \qquad (3.7)$$

In contrast to RRC and SwRC where the cooperation gain grows roughly linearly with the number of relays, the cooperation gain for the StRC strategy is dichotomous: Below the threshold $n_J C_J$, there is no cooperation gain; once the number of relay nodes passes the threshold, the cooperation gain rises dramatically, especially for the ARxC receivers. For instance, given $C = 256$, $p_J = 0.2$, and $n_J C_J = n_s C_s$, the cooperation gain with $n = 40$ relay nodes is 115.2 for StRC with the adaptive receiver, approximately 2.9 times greater than RRC or SwRC. Meanwhile, it is as small as 19 for StRC with the RRxC receiver.

If two or more transmissions on the same channel lead to a failure in reception due to the difference in the arrival time of transmitted packets, we have the follow:

Lemma 3.5. *The successful packet reception rate of the collaborative UFH-based broadcast with RRC and RRxC strategies without perfect synchronization, denoted by $p_a^{RRCAsyn}$, is given by*

$$p_a^{RRCAsyn}(n) = (n+1)\frac{1}{C}\left(1-\frac{1}{C}\right)^n(1-p_J). \tag{3.8}$$

Proof. Each source or relay node under the RRC strategy transmits over a given channel with a probability $1/C$. Hence the probability for exactly one out of these $n+1$ transmitters to work on a given channel (i.e., one node selects that channel and all the other n nodes choose other channels) can be written as $(n+1)(1-1/C)^n/C$. Without perfect synchronization, the receiver can obtain the packet, if working on a channel that is clear from jamming with probability of $1-p_J$, and is selected by exactly one of the $n+1$ transmitters. Thus the successful packet reception rate is given by $(n+1)(1-1/C)^n(1-p_J)/C$.

As shown in Eq. (3.8), the successful packet reception rate for the RRC strategy is proportional to $n+1$, because more packet copies are provided to the receiver by more relay nodes. On the other hand, the factor $(1-1/C)^n$ decreases with n, which accounts for more channel collision happening with more relay nodes.

Lemma 3.6. *The successful packet reception rate of the collaborative UFH-based broadcast with SwRC and RRxC strategies without perfect synchronization, denoted by $p_a^{SwRCAsyn}$, is given by*

$$p_a^{SwRCAsyn}(n) = \left(\frac{1}{C}\left(1-\frac{n}{C}\right)+\left(1-\frac{1}{C}\right)\frac{n}{C}\right)(1-p_J). \tag{3.9}$$

Proof. See [5].

Lemma 3.7. *The successful packet reception rate of the collaborative UFH-based broadcast with StRC and RRxC strategies without perfect synchronization, denoted by $p_{RRxC,a}^{StRCAsyn}$, is given by*

$$p_{RRxC,a}^{StRCAsyn}(n) = \frac{n}{C}\left(1-\frac{n_J C_J}{C}\right)\left(1-\frac{1}{C}\right)+\left(1-\frac{n}{C}\right)\frac{1}{C}, \tag{3.10}$$

and that with StRC and ARxC is given by

$$p_{ARxC,a}^{StRCAsyn}(n) = \left(1-\frac{n_J C_J}{C}\right)\left(1-\frac{1}{C}\right). \tag{3.11}$$

Proof. See [5].

If C is large, the synchronization error only leads to small performance degradation in the collaborative broadcast. In addition, the successful packet reception rate for the collaborative broadcast mostly increases with the number of relay nodes.

We now evaluate the communication efficiency of the collaborative UFH-based broadcast in large-scale networks under a large number of nodes n, which helps reveal the scalability of the broadcast regarding the network size and other system parameters, such as the data rate of a channel denoted by R. A message with L bits is divided into M short packets to transmit separately. Each packet is expanded with O-bit overhead including the message ID, fragment number, hash index, etc, based on the same modulation scheme. In the slotted and ideally synchronized network, the broadcast duration of a packet, denoted by T_s, is given by

$$T_s = \frac{O + \frac{L}{M}}{R}. \tag{3.12}$$

The network has to control the number of relays, denoted by n_r. Intuitively, allowing more relays than the number of available channels in the network degrades the broadcast performance, because the node collisions due to simultaneous transmissions yield serious network congestions. On the other hand, a small number of relay nodes results in inadequate collaboration gain. The packet reception rate against the jammer is given by:

$$p_{n_r}(M) = M\left(\left(1 - \left(1 - \frac{1}{M}\right)\frac{1}{C}\right)^{n_r} - \left(1 - \frac{1}{C}\right)^{n_r}\right)(1 - P_{jam}), \tag{3.13}$$

where $P_{jam} = C_j T_s / C$, and C_j is the number of channels that the jammer can efficiently block in a time slot. The optimal number of the relay nodes is given by the following lemma:

Lemma 3.8. *The optimal numbers of transmitters, in terms of maximizing packet reception rate $p_{n_r}(M)$, is given by*

$$n_r^*(M) \approx -M \ln\left(1 - \frac{1}{M}\right)C. \tag{3.14}$$

Proof. See [9].

The performance loss is trivial, if the number of transmitters is within the range $[C, 1.4C]$. Due to the dynamic relay accumulation, it is challenging to derive the exact expression of the average network broadcast delay, denoted by D^c. Instead, we explore its lower bound. The process is accelerated after the first relay appears. Before that no cooperation can be exploited and there is no difference between CUFH and UFH. Let $D_{\min}^c$ denote the time duration from the beginning of the broadcast to the first relay appearing, which contributes a substantial portion in the total network broadcast delay. Thus $E\left(D_{\min}^c\right)$ serves as a lower bound for $E(D^c)$, and we have

$$E\left(D^c\right) \geq E\left(D_{\min}^c\right) > \sum_{i=0}^{\infty}\left(1-\left(1-\left(1-\frac{1}{C}\left(1-P_{jam}\right)\right)^i\right)^M\right)^n, \qquad (3.15)$$

where $D^c = \max_{1 \leq i \leq n}\left(D_i^c\right) \geq D_{\min}^c = \min_{1 \leq i \leq n}\left(D_i^c\right)$, and D_i^c is the broadcast delay for node i.

CUFH-p proposed in [9] as an improved version of CUFH for large-scale networks can be easily implemented in practice. The source node predetermines $n_r - 1$ destination nodes as potential relays and imbeds their IDs in the message broadcast. Then the source transmits the packets sequentially using UFH. These predetermined relay nodes serve as relays right after obtaining the whole message. The relays randomly and independently select channels for the transmission of each packet.

Lemma 3.9. *The average network broadcast delay of CUFH-p is given by*

$$E\left(D^p\right) < \sum_{i=0}^{\infty}\left(1-\varepsilon(n_r,i)^{n-n_r+1}\right) + \frac{C}{1-P_{jam}}\sum_{k=1}^{M(n_r-1)}\frac{1}{k} + 1, \qquad (3.16)$$

where D^p is network broadcast delay that CUFH incurs and $\varepsilon(n_r, i)$ is given in Eq. (13) in [9].

Proof. See [9].

The relay nodes in CUFH-p are predetermined rather than dynamically selected among all the destination nodes as in CUFH, resulting in some performance degradation, since non-relay nodes cannot help even if they obtain the message before some relay nodes. The fixed relay selection significantly simplifies the implementation of CUFH, as dynamical constraint on the number of relays to a predetermined value usually requires feedbacks from destination nodes to the source node. Analysis and simulation results show that CUFH-p outperforms UFH significantly, and asymptotically achieves the optimal cooperation gain [9].

3.5 Implementation Issues

In the collaborative broadcast based on UFH, each transmitter, either the source node or a relay, repeatedly sends the broadcast packets and stops the transmission once receiving the ACKs from all its neighboring nodes or reaching the maximum transmission duration, denoted by Δ, whichever comes first. As shown in Fig. 3.6, each relay node starts and stops transmission at different time, with the transmission

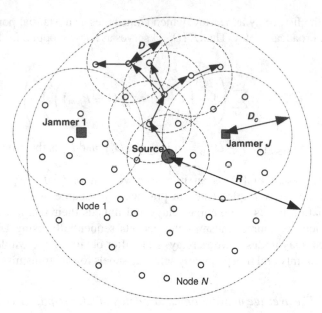

Fig. 3.6 Illustration of the collaborative broadcast in multi-hop cognitive radio networks with N receivers against J jammers with the coverage range denoted by D_c, in which each relay node starts and stops transmission at different time with the transmission range denoted by D

range of the relay (or jammer) denoted by D (or D_c). The limit on the transmission duration is introduced to address the possible loss of the ACK signals due to radio channel imperfection, packet collisions, or attacks by the jammers.

Each ACK message, including the message ID, receiver ID, and time stamp, is sent on a fixed and known channel by a node right after successfully obtaining all M packets. An authentication mechanism for the ACK signals addresses spoofing by jammers. Local interference (including intentional jamming) on this common channel can be detected, and further actions can be taken accordingly. If the ACK mechanism fails, it resorts to the time-out mechanism to control the transmission duration, in which Δ is a key parameter based on the estimate of the average broadcast delay with the RRxC strategy.

Assume that a transmitter periodically broadcasts M packets to l nodes within its communication range. It is clear that the probability for all these l independent receivers to obtain all the M packets during the first m slots is $P[m] = (1 - (1 - p_a)^m)^{Ml}$, where $p_a = (1 - p_J)/C$ is the successful packet reception rate of a receiver. Following the analysis in [8], the average broadcast delay in terms of time slots is given by

$$T_{avg}^{hop} = M \sum_{m=0}^{\infty} \left[1 - (1 - (1 - p_a)^m)^{Ml} \right]. \tag{3.17}$$

In a single-hop network, $l = N$, and N is the number of nodes to receive the message. In a multi-hop setting with uniform node placement, the average number of the nodes within the reach of one hop is given by $l = D^2 N / R^2$, where D is the signal coverage radius and R is the radius of the network. Finally, the transmission duration Δ can be set as

$$\Delta = a T_{avg}^{hop} = a M \sum_{m=0}^{\infty} \left[1 - (1 - (1 - p_a)^m)^{Ml} \right], \qquad (3.18)$$

where the constant a can be fine tuned in practice. It should be noted that Δ depends on the jamming probability, which could potentially be exploited by intelligent jammers. Hence Δ should ideally be updated during the broadcast process, and this issue deserves further study. It is found through simulation that the system performance is not sensitive to the choice of Δ, and in practice the parameter may be adjusted according to the need (e.g., set to a larger value if jamming is particularly a concern).

In the broadcast based on RRC, each node other than the source node first enters the receiving mode, in which a node independently and randomly selects one out of C channels and listens, and switches to another randomly selected channel after one or several time slots to counteract jamming. This process repeats until the node successfully receives all M packets. Next, the node informs its neighbors about this information with an ACK signal, which contains the message ID, node ID, and time stamp, and is sent on a fixed and known channel.

The transmission modes of different nodes start at different times, e.g., the source node enters this mode from the very beginning while a node at the edge of the network may never enter the transmission mode. In the transmission mode, each node randomly selects a channel out of the C channels and sends a packet. In order to deal with the possible loss of ACK signals due to channel imperfection or jamming, each transmission stops after Δ slots, given by Eq. (3.18), even without receiving enough ACK signals. The node repeats this process to send all M packets in sequence, until it receives all the ACK signals from its neighbors, or Δ time slots elapse, whichever comes first.

After having successfully received all M packets, each node with StRC relays the message on a fixed channel, that is assumed to be distinctly related to its unique node ID. Each node is assumed to know the relay channels that its neighbors may use. In order to counteract smart jamming, each node has two receiving modes, based on whether any relay channel is not blocked: If that is true, the node focuses on the relay channels by randomly selecting one of the potential relay channels in the neighborhood; otherwise, the node randomly selects one out of all the C channels.

In the broadcast based on StRC, the receiver sets a status flag to be true at the beginning, and updates it according to the checking results of recently received packets. When working on the relay channels, the receiver changes the flag to be false, if failing to receive all the recent R_p packets, which means that all these relay channels are very likely to be jammed. The parameter R_p can be set as the actual

---

Algorithm: Collaborative Anti-jamming Broadcast with RRC

While The node has not received all M packets yet **do**

    ChID = an integer randomly selected from $[1, C]$;

    Listen to the *ChID*-th channel;

end

Send ACK (Message ID, Node ID, Time Stamp);

$\Delta \leftarrow Eq. (3.18)$;

for $i \leftarrow 1$ *to* Δ **do**

    if Not all the ACKs from the neighbors are received

    then

        ChID = an integer randomly selected in $[1, C]$;

        Send a packet sequentially on the *ChID*-th channel;

    else

        Stop transmission;

    end

end

---

number of neighboring nodes, or the average number of neighboring nodes if the former is unknown. When coming across a clear relay channel, the node sets the flag to be true and focuses on the relay channels again. After receiving all the packets, the node sends an ACK signal to its neighbors and enters to the transmission mode. Then the node sends the packets on a fixed channel corresponding to its preassigned unique node ID. The transmission duration is also controlled by a timer of length Δ.

The MAC strategies can achieve the minimal broadcast delay or significantly reduce the overall energy consumption in the UFH-based anti-jamming broadcast without pre-shared keys was proposed in [10]. The overall energy consumption consists of the total transmission energy consumed by the transmitters including both the source node and relay nodes, and the energy consumed by the receivers. Denote the energy consumption for a node to send and to receive a packet as E_t and E_r respectively. The average number of transmitters and receivers during a time slot are given by $N_t = 1 + np$ and $N_t = N - n$, respectively, and $1/p_a$ represents the average number of transmissions for a successful packet reception. Therefore, the effective energy consumption is given by

$$E_{eff} \triangleq \frac{E_t N_t + E_r N_r}{p_a} = \frac{E_t (1 + np) + E_r (N - n)}{p_a}. \tag{3.19}$$

Lemma 3.10. *If $C \gg 1$ and $n < C$, the access probability of the collaborative UFH-based broadcast with perfect synchronization can be well approximated by*

$$p^* \approx \min\left(1, \frac{\sqrt{n^2\left(1 + \frac{E_r}{E_t}(N-n)\right)^2 + 4C(n^2 - n)\left(1 + \frac{E_r}{E_t}(N-n)\right)} - n\left(1 + \frac{E_r}{E_t}(N-n)\right)}{-2C\left(1 + \frac{E_r}{E_t}(N-n)\right)}\right). \tag{3.20}$$

Proof. See [10].

Algorithm: Collaborative Anti-jamming Broadcast with StRC

FlgClearRelayChannel = True

While The node has not received all *M* packets yet **do**

    if *FlgClearRelayChannel* = *True* **then**

        ChID = an integer randomly selected from the relay channel set in its neighborhood;

    else

        ChID = an integer randomly selected from [1, *C*];

    end

    Listen to the *ChID* − *th* channel;

    if *FlgClearRelayChannel* = *True* **then**

        if All recent R_p packets are jammed **then**

            FlgClearRelayChannel = *False*;

        end

    else

        if *ChID*-th channel is a unblocked relay channel **then**

            FlgClearRelayChannel = *True*;

        end

    end

end

Send ACK (Message ID, Node ID, Time Stamp);

$\Delta \leftarrow$ *Eq.* (3.18);

ChID = an integer derived from its Node ID;

for $i \leftarrow 1$ *to* Δ **do**

    if Not all the ACKs from the neighbors are received **then**

        Send a packet on the *ChID*-th channel;

    else

        Stop transmission;

    end

end

3.6 Summary

In this chapter, we have presented the anti-jamming transmissions based on uncoordinated spread spectrum, focusing on the collaborative UFH-based broadcast that applies node cooperation to exploit the frequency and spatial diversities to enhance the communication efficiency against smart jammers. The random relay channel selection in the broadcast provides a cooperation gain proportional to the number of relay nodes, and is amenable to simple distributed implementation. The static relay selection strategy substantially further improves the cooperation gain under weak jamming relative to the collaboration scale. We have examined the communication efficiency of UFH and the collaborative broadcast, showing that the broadcast is robust against the relay synchronization error. We have investigated the implementation issues for the collaborative broadcast with uncoordinated FH and DSSS, such as the ACK and time-out mechanisms.

References

1. Strasser, M., Pöpper, C., Čapkun, S.: Efficient uncoordinated FHSS anti-jamming communication. In: Proceedings of ACM International Symposium on Mobile Ad Hoc Networking and Computing (MobiHoc), pp. 207–218 (2009)
2. Popper, C., Strasser, M., Capkun, S.: Anti-jamming broadcast communication using uncoordinated spread spectrum techniques. IEEE J. Sel. Areas Commun. **28**(5), 703–715 (2010)
3. Strasser, M., Capkun, S., Cagalj, M.: Jamming-resistant key establishment using uncoordinated frequency hopping. In: Proceedings of IEEE Symposium on Security and Privacy, pp. 64–78 (2008)
4. Liu, A., Ning, P., Dai, H., Liu, Y., Wang, C.: USD-FH: Jamming-resistant wireless communication using frequency hopping with uncoordinated seed disclosure. In: Proceedings of IEEE International Conference on Mobile Ad hoc and Sensor Systems (MASS), pp. 41–50 (2010)
5. Xiao, L., Dai, H., Ning, P.: Jamming-resistant collaborative broadcast using uncoordinated frequency hopping. IEEE Trans. Inf. Forensics Secur. **7**(1), 297–309 (2012)
6. Pöpper, C., Strasser, M., Capkun, S.: Jamming-resistant broadcast communication without shared keys. In: Proceedings of USENIX Security Symposium, pp. 231–248 (2009)
7. Gilbert, S., Guerraoui, R., Kowalski, D.R., Newport, C.: Interference-resilient information exchange. In: Proceedings of IEEE International Conference on Computer Communications (INFOCOM), pp. 2249–2257 (2009)
8. Slater, D., Tague, P., Poovendran, R., Matt, B.J.: A coding-theoretic approach for efficient message verification over insecure channels. In: Proceedings of ACM Conference on Wireless Network Security, pp. 151–160 (2009)
9. Li, C., Dai, H., Xiao, L., Ning, P.: Communication efficiency of anti-jamming broadcast in large-scale multi-channel wireless networks. IEEE Trans. Signal Process. **60**(10), 5281–5292 (2012)
10. Xiao, L., Dai, H., Ning, P.: MAC design of uncoordinated FH-based collaborative broadcast. IEEE Trans. Wirel. Commun. Lett. **1**(3), 261–264 (2012)

Chapter 4
Game Theoretic Study on Jamming in CRNs

4.1 Introduction

As a powerful tool in strategic decision making, game theory has shown strength to address jamming attacks in wireless communications [1]. For example, the Nash equilibrium (NE) of a zero-sum jamming game in cognitive radio networks with perfect channel information was investigated in [2]. Anti-jamming games with inaccurate knowledge on the number of attackers and environmental parameters were analyzed in [3]. In [4], a stochastic jamming game was formulated for multi-carrier CRNs. A non-cooperative random access game was used to investigate jamming attacks with unknown system parameters [5]. A zero-sum game with unknown jamming strategy for sub-carriers and fading channel gains was investigated in [6], while a nonzero-sum matrix game with inaccurate histories of the opponents was analyzed in [7]. The saddle-point strategy of a dynamic zero-sum game with asymmetric information between a transmitter and jammer was provided in [8].

Smart jammers can learn the transmission patterns to determine their jamming strategies and thus throw serious threats to wireless networks [9]. Being able to describe the hierarchical behaviors among players, Stackelberg game [10, 11] provides a powerful method to address smart jammers. For instance, the jamming power control was formulated as a Stackelberg game in [9], in which the jamming power is chosen according to the ongoing transmit power. In addition, the Stackelberg equilibrium (SE) of a game between a primary user and a secondary user was presented in [12].

The power control for cooperative communication networks was formulated as cooperative games [13]. In the anti-jamming cooperative wireless networks investigated in [11], a source node and a relay node send a same message on a single channel to resist smart jammers.

Traditional decision-making models in game theory are based on the expected utility of rational players without any uncertainty or error. In [14], prospect theory

L. Xiao, *Anti-Jamming Transmissions in Cognitive Radio Networks*,
SpringerBriefs in Electrical and Computer Engineering,
DOI 10.1007/978-3-319-24292-7_4

was applied to investigate the anti-jamming communications in wireless networks, where users are subjective in decision-making under uncertain environments.

4.2 Static Jamming Games

We first consider a static jamming game, denoted by $\mathbf{G}^1$, in which a jammer and a second user choose their transmit power independently in a time slot. In this game, the transmit power of the SU is $P_s \in [0, P_s^{max}]$, and the jamming power of the jammer is $P_j \in [0, P_j^{max}]$, where P_x^{max} is the maximum power of Player x, with $x = s$ (i.e., SU) or j (i.e., jammer). Let h_s (or h_j) denote the channel power gain between the transmitter (or jammer) and the legitimate receiver, C_x be the transmission cost per unit power of Player x, and σ be the noise power. The presence indicator of PUs is denoted as α, which equals one in the absence of PU and zero otherwise. Both players know these system parameters, possibly by learning from the interaction history.

The SU (or jammer) takes action P_s (or P_j) to maximize its individual utility, denoted by U_s (or U_j). Based on the SINR of the signals of the SU, we have

$$U_s\left(P_s, P_j\right) = \frac{h_s P_s \alpha}{\sigma + h_j P_j} - C_s P_s, \tag{4.1}$$

$$U_j\left(P_s, P_j\right) = -\frac{h_s P_s \alpha}{\sigma + h_j P_j} + C_s P_s - C_j P_j, \tag{4.2}$$

where $C_s P_s$ indicates that the jammer aims to deplete the SU's battery level, and α denotes the priority of PU. In the presence of a PU with $\alpha = 0$, as both U_s and U_j decrease with P_s and P_j, both the SU and the jammer stop transmitting. Otherwise, the SU and the jammer compete for higher individual utilities.

4.2.1 Nash Equilibrium

Each NE of a static jamming game consists of the optimal strategies that neither the jammer nor user can improve its utility by changing its power unilaterally, i.e., $\forall 0 \leq P_s \leq P_s^{max}$ and $0 \leq P_j \leq P_j^{max}$, we have

$$U_s\left(P_s^{NE}, P_j^{NE}\right) \geq U_s\left(P_s, P_j^{NE}\right), \tag{4.3}$$

$$U_j\left(P_s^{NE}, P_j^{NE}\right) \geq U_j\left(P_s^{NE}, P_j\right). \tag{4.4}$$

Lemma 4.1. *The NE of the static anti-jamming game* $\mathbf{G}^1$ *is given by*

$$
\left(P_s^{NE}, P_j^{NE}\right) = \begin{cases}
(0,0)\,, & I_1 \\[2mm]
\left(\frac{C_j h_s}{C_s^2 h_j}, \frac{1}{h_j}\left(\frac{h_s}{C_s} - \sigma\right)\right), & I_2 \\[2mm]
\left(P_s^{max}, P_j^{max}\right), & I_3 \\[2mm]
\left(P_s^{max}, 0\right), & I_4 \\[2mm]
\left(P_s^{max}, \frac{1}{h_j}\left(\sqrt{\frac{h_s h_j P_s^{max}}{C_j}} - \sigma\right)\right), & I_5,
\end{cases}
\tag{4.5}
$$

where the conditions are given by

$$
I_1 : \alpha = 0, \text{ or } C_s > \frac{h_s}{\sigma};
$$

$$
I_2 : \alpha = 1, \frac{h_s}{P_j^{max} h_j + \sigma} \le C_s \le \frac{h_s}{\sigma}, \; C_j \le \frac{C_s^2 P_s^{max} h_j}{h_s};
$$

$$
I_3 : \alpha = 1, \; C_s < \frac{h_s}{P_j^{max} h_j + \sigma}, \; C_j < \frac{h_s P_s^{max} h_j}{\left(P_j^{max} h_j + \sigma\right)^2};
$$

$$
I_4 : \alpha = 1, \; C_s < \frac{h_s}{P_j^{max} h_j + \sigma}, \; C_j > \frac{h_s P_s^{max} h_j}{\sigma^2},
$$

$$
\text{or } \frac{h_s}{P_j^{max} h_j + \sigma} \le C_s \le \min\left(\frac{h_s}{\sigma}, \sqrt{\frac{C_j h_s}{P_s^{max} h_j}}\right);
$$

$$
I_5 : \alpha = 1, \; C_s < \frac{h_s}{P_j^{max} h_j + \sigma}, \; \frac{h_s P_s^{max} h_j}{\left(P_j^{max} h_j + \sigma\right)^2} \le C_j \le \frac{h_s P_s^{max} h_j}{\sigma^2}.
$$

Proof. See [10].

If a PU occupies the channel (i.e., $\alpha = 0$), or the transmit cost of the SU is large (i.e., $C_s > \frac{h_s}{\sigma}$), the SU stops transmitting, leading to the silence of the jammer. However, if both the transmit cost of the SU and the jamming cost are very small, i.e., condition I_3, both the SU and the jammer apply their highest transmission and jamming power. The transmit power and jamming power depends on the channel conditions (i.e., h_s and h_j) and energy costs (i.e., C_s and C_j), if the jammer has a low jamming cost and the transmit cost of the SU C_s is in the range of $\left[\frac{h_s}{P_j^{max} h_j + \sigma}, \frac{h_s}{\sigma}\right]$ (i.e., condition I_2). The SU applies its maximum transmit power if the cost C_s is small, (i.e., conditions I_3, I_4 and I_5). Neither the user nor jammer can improve its utility by unilaterally deviating from the NE strategy of the game.

4.2.2   Stackelberg Equilibrium

The hierarchical interactions between a jammer and an SU can be modeled as a Stackelberg game, in which the SU as the leader first transmits with power $P_s \in [0, P_s^{max}]$, and then the smart jammer as the follower chooses its jamming power $P_j \in [0, P_j^{max}]$.

The Stackelberg anti-jamming game, denoted by $\mathbf{G}^2 = \langle \{s, j\}, \{P_s, P_j\}, \{U_s, U_j\} \rangle$, consists of an SU as the leader and a jammer as the follower, both choosing their transmit power according to the channel gains and transmission costs. The utilities of the players, U_s and U_j, are defined in (4.1) and (4.2), respectively. The relative observation error of the jammer, denoted by ϵ, is defined as $\epsilon = |\hat{P}_s - P_s|/P_s$, where $\hat{P}_s$ represents the observed transmit power of the SU.

The jammer is assumed to ignore its unknown observation error when choosing its jamming power to maximize its expected utility, which is obtained by replacing P_s in (4.2) with $\hat{P}_s$. Thus, the utility expected by the jammer, denoted by $\hat{U}_j$, is

$$\hat{U}_j \left(\hat{P}_s, P_j \right) = -\frac{h_s \hat{P}_s \alpha}{\sigma + h_j P_j} + C_s \hat{P}_s - C_j P_j. \qquad (4.6)$$

Meanwhile, as the leader, the SU assumes that the jammer can observe its accurate transmit power, and decides its transmit power by estimating the jamming strategy $\hat{P}_j$ to substitute $\hat{P}_j$ in (4.1). Thus, the estimated utility of the SU, denoted by $\hat{U}_s$, is given by

$$\hat{U}_s \left(P_s, \hat{P}_j \right) = \frac{h_s P_s \alpha}{\sigma + h_j \hat{P}_j} - C_s P_s. \qquad (4.7)$$

The Stackelberg equilibrium of the game $\mathbf{G}^2$ under the power constraints, denoted by $\left(P_s^{SE}, P_j^{SE} \right)$, is given by

$$P_s^{SE} = \arg \max_{0 \leq P_s \leq P_s^{max}} \hat{U}_s \left(P_s, \hat{P}_j \right) \qquad (4.8)$$

$$P_j^{SE} = \arg \max_{0 \leq P_j \leq P_j^{max}} \hat{U}_j \left(\hat{P}_s, P_j \right). \qquad (4.9)$$

Therefore, we have the following:

Lemma 4.2. *The SE of the anti-jamming game* $\mathbf{G}^2$, $\left(P_j^{SE}, P_s^{SE} \right)$ *is given by*

$$P_j^{SE} = \begin{cases} 0, & \alpha = 0, \text{ or } \hat{P}_s \leq \frac{\sigma^2 C_j}{h_s h_j} \\ P_j^{max}, & \alpha = 1, \hat{P}_s \geq \frac{C_j (P_j^{max} h_j + \sigma)^2}{h_s h_j} \\ \frac{1}{h_j} \left(\sqrt{\frac{h_s h_j \hat{P}_s}{C_j}} - \sigma \right), & \text{otherwise.} \end{cases} \qquad (4.10)$$

$$P_s^{SE} = \begin{cases} 0, & \Pi_1 \\ \frac{h_s C_j}{4 h_j C_s^2}, & \Pi_2 \\ \frac{\sigma^2 C_j}{h_s h_j}, & \Pi_3 \\ P_s^{max}, & o.w., \end{cases} \tag{4.11}$$

where

$\Pi_1 : \alpha = 0, \text{ or } C_s > \dfrac{h_s}{\sigma};$

$\Pi_2 : \alpha = 1, \dfrac{h_s}{P_j^{max} h_j + \sigma} \le C_s < \dfrac{h_s}{2\sigma}, \; C_j < \dfrac{4 h_j C_s^2 P_s^{max}}{h_s},$

$\quad or \; \alpha = 1, \; C_s < \min\left(\dfrac{h_s}{2\sigma}, \dfrac{h_s}{P_j^{max} h_j + \sigma}\right), \; C_j < \dfrac{4 h_j C_s^2 P_s^{max}}{h_s},$

$\quad or \; \alpha = 1, \; C_s < \min\left(\dfrac{h_s}{2\sigma}, \dfrac{h_s}{P_j^{max} h_j + \sigma}\right),$

$\quad \dfrac{4 P_s^{max} h_j C_s}{h_s}\left(\dfrac{h_s}{P_j^{max} h_j + \sigma} - C_s\right) < C_j < \dfrac{4 P_s^{max} C_s^2 h_j}{h_s};$

$\Pi_3 : \alpha = 1, \; \max\left(\dfrac{h_s}{2\sigma}, \dfrac{h_s}{P_j^{max} h_j + \sigma}\right) \le C_s \le \dfrac{h_s}{\sigma}, \; C_j < \dfrac{P_s^{max} h_s h_j}{\sigma^2},$

$\quad or \; \alpha = 1, \; \dfrac{h_s}{2\sigma} < C_s < \dfrac{h_s}{P_j^{max} h_j + \sigma},$

$\quad \dfrac{P_s^{max} h_s h_j}{h_s \sigma - \sigma^2 C_s}\left(\dfrac{h_s}{P_j^{max} h_j + \sigma} - C_s\right) < C_j < \dfrac{P_s^{max} h_s h_j}{\sigma^2};$

Proof. See [10].

If the jammer finds low transmission power of the SU, indicating that the SU's transmission rate is low compared with the jamming cost, $\hat{P}_s \le \sigma^2 C_j/(h_s h_j)$, the optimal jamming strategy is to ignore the current transmission. As another extreme case, if the jamming cost is negligible compared with the ongoing transmission, i.e., $\hat{P}_s \ge C_j \left(P_j^{max} h_j + \sigma\right)^2/(h_s h_j)$, the optimal strategy is to apply the highest jamming power to block the ongoing transmission. Otherwise, if

$$\frac{\sigma^2 C_j}{h_s h_j} < \hat{P}_s < \frac{C_j \left(P_j^{max} h_j + \sigma\right)^2}{h_s h_j}, \tag{4.12}$$

the optimal strategy is to adjust the jamming power according to the observed transmit power. If the transmission cost is high enough (i.e., $C_s \geq h_s/\sigma$), the optimal SU's strategy is to stop transmitting. As another extreme case with a small transmission cost compared with the jamming cost and channel conditions (i.e., the last condition), the optimal SU's strategy is to maximize its power to increase the SINR. If the transmission cost is relatively small compared with the channel conditions (i.e., condition Π_3), the transmit power of the SU mostly depends on the channel conditions instead of C_s. Both the SU and the jammer stop transmitting in the presence of PUs in order to avoid interfering with the latter.

4.3   Cooperative Anti-jamming Games

We consider a cooperative wireless network against a smart jammer, where a relay node helps the source broadcast a message on a single channel to the destination. The source, relay node, and jammers can flexibly choose their transmission power levels to maximize their individual utilities. The relay node decides its transmission power after knowing the source's transmission power. On the other hand, once sensing the power of the source and relay node, the smart jammer chooses its jamming power on the channel to maximize the damages.

In the Stackelberg game, the players take actions sequentially to control the power. The interactions among the legitimate users and the jammer are modeled as a Stackelberg game, denoted by $\mathbf{G}^3$. As the reader in the game the source node first choose its transmit power. The relay determines its power after the leader. As a follower in the game, the jammer is the last to choose its transmit power. Let $x = s, r$ and j denote the source, the relay node and the jammer, respectively. The action of Player x is its own transmission power denoted by $P_x \in [0, \infty)$. Let $h_x > 0$ denote the channel gain between Player x and the receiver, and $C_x > 0$ be the transmission cost per unit power of Player x. The SINR of the signal at the receiver is given by

$$SINR = \frac{h_s P_s + h_r P_r}{N + h_j P_j}, \tag{4.13}$$

where N is the noise power.

The utility of the source node (or relay node), denoted by U_s (or U_r), is based on the SINR and the transmit cost, given by

$$U_s \left(P_s, P_r, P_j\right) = \frac{h_s P_s + h_r P_r}{N + h_j P_j} - C_s P_s, \tag{4.14}$$

$$U_r(P_s, P_r, P_j) = \frac{h_s P_s + h_r P_r}{N + h_j P_j} - C_r P_r. \tag{4.15}$$

The jammer aims to block the ongoing transmission with less jamming costs and waste more power of the source node and relay. The SINR-based utility of the jammer, denoted by U_j, is given by

$$U_j(P_s, P_r, P_j) = -\frac{h_s P_s + h_r P_r}{N + h_j P_j} + C_s P_s + C_r P_r - C_j P_j. \tag{4.16}$$

In this Stackelberg game, each player chooses its strategy in sequence to maximize its individual utilities. Following the power constraints, the SE of the game $\mathbf{G}^3$, denoted by $\left(P_s^{SE}, P_r^{SE}, P_j^{SE}\right)$, are given by

$$P_s^{SE} = \arg\max_{P_s \geq 0} U_s(P_s, \hat{P}_r, \hat{P}_j), \tag{4.17}$$

$$P_r^{SE} = \arg\max_{P_r \geq 0} U_r(\hat{P}_s, P_r, \hat{P}_j), \tag{4.18}$$

$$P_j^{SE} = \arg\max_{P_j \geq 0} U_j(\hat{P}_s, \hat{P}_r, P_j), \tag{4.19}$$

where $\hat{P}_x$ is the transmit power of Player x forecasted by the other player. Based on the location information of the other players, each can estimate the other's transmission power accurately (i.e., $\hat{P}_x = P_x$).

The game between the transmitters and the jammer involves three optimization sub-problems. In the Stackelberg game, the source first chooses a strategy to maximize its utility U_s. Then the relay node chooses its power to achieve the maximum its utility U_r. Finally, after observing the strategy of the transmitters, the jammer chooses its jamming power to maximize its utility U_j. The Stackelberg equilibrium consists of the optimal strategies of the three players. In order to derive the Stackelberg equilibrium of the jamming game $\mathbf{G}^3$, denoted by $\left(P_s^{SE}, P_r^{SE}, P_j^{SE}\right)$, we analyze the impact of the source node's power on the jammer and the relay node.

Lemma 4.3. *The SE of the jamming game* $\mathbf{G}^3$ *is given by:*

$$P_s^{SE} = \begin{cases} \frac{h_s C_j}{4 h_j C_s^2}, & \Phi \\ \frac{C_j N^2}{h_s h_j}, & C_r \geq \frac{h_r}{N}, \frac{h_s}{2N} \leq C_s < \frac{h_s}{N} \\ 0, & \text{otherwise,} \end{cases} \tag{4.20}$$

where $\Phi : C_r \leq \frac{h_r}{2N}, \frac{C_s}{C_r} < \frac{h_s}{2h_r}$ *or* $\frac{h_r}{2N} < C_r < \frac{h_r}{N}, C_s < \frac{h_s}{4N}$ *or* $C_r \geq \frac{h_r}{N}, C_s < \frac{h_s}{2N}$;

$$P_r^{SE} = \begin{cases} 0, & \Omega_1 \\ \frac{h_r C_j}{4 h_j C_r^2} - \frac{h_s P_s}{h_r}, & \Omega_2 \\ \frac{1}{h_r}\left(\frac{C_j N^2}{h_j} - h_s P_s\right), & \text{otherwise,} \end{cases} \tag{4.21}$$

where

$$\Omega_1 : P_s \geq \max\left(\frac{C_j N^2}{h_s h_j}, \frac{h_R^2 C_j}{4 h_s h_j C_r^2}\right) \text{ or } P_s < \frac{C_j N^2}{h_s h_j}, \frac{h_r}{N} \leq C_r,$$

$$\Omega_2 : \frac{C_j N^2}{h_s h_j} \leq P_s < \frac{h_r^2 C_j}{4 h_s h_j C_r^2} \text{ or } P_s < \frac{C_j N^2}{h_s h_j}, \frac{h_r}{N} \geq 2 C_r;$$

and

$$P_j^{SE} = \begin{cases} 0, & h_s P_s + h_r P_r \leq \frac{C_j N^2}{h_j} \\ \frac{1}{h_j}\left[\sqrt{\frac{h_j(h_s P_s + h_r P_r)}{C_j}} - N\right], & \text{otherwise.} \end{cases} \tag{4.22}$$

Proof. See [11].

Based on (4.20), the source as the leader chooses its power based on the relay and jamming the power, as well as the channel conditions. If the transmission costs of the source and relay node are both too high (i.e., $C_r \geq \frac{h_r}{N}$ and $C_s \geq \frac{h_s}{N}$) or the transmission gain of the relay is high, the source's optimal strategy is to stop transmission. Otherwise, the best strategy of the source is to adjust its power based on all channel gains and the jamming cost. If the transmit power of the source known by the relay is high enough (i.e., $P_s \geq \max\left(\frac{C_j N^2}{h_s h_j}, \frac{h_r^2 C_j}{4 h_s h_j C_r^2}\right)$), the optimal relay strategy is to stop the transmit. If the power of the source and the transmission cost of the relay are both low (i.e., $P_s < \frac{C_j N^2}{h_s h_j}$ and $\frac{h_r}{N} \leq C_r$), the relay cannot help the source. Otherwise, the relay chooses its power based on the current transmission power of the source. According to (4.22), the optimal jamming strategy varies with the ongoing transmission power. If the jamming gain is less than the jamming cost due to the low ongoing transmission power (i.e., $h_s P_s + h_r P_r \leq \frac{C_j N^2}{h_j}$), the jammer's best strategy is to ignore the present transmission. However, once the current transmission power exceeds a certain threshold (i.e., $h_s P_s + h_r P_r > \frac{C_j N^2}{h_j}$), the optimal jamming strategy is to change according to the current transmission power of the source and the relay.

4.4 Repeated Jamming Games

The interactions between a legal transmitter and a jammer in cognitive radio networks over multiple time slots can be formulated as a repeated jamming game. In the Stackelberg game $\mathbf{G}^4 = \langle \{s, j\}, \{P_s, P_j\}, \{U_s, U_j\} \rangle$, the SU and the jammer choose their transmit power to maximize the long-term utilities. Markov decision process is commonly used to model the jamming process in dynamic games. The jamming power and the transmit power are quantized into discrete levels. The action of the jammer is denoted by $0 \leq s \leq S$, where S is the jamming power constraint. The state transition probability of the jammer, denoted by $p(s'|s)$, represents the probability that the jamming power will change from s at time k to s' at time $k + 1$, with $s, s' \in \{0, 1, 2, \ldots, S\}$.

In the repeated jamming game, the channel gain is time variant and can be modeled with a Markov chain. More specifically, the probability for the channel gain to change from h at time slot k to h' at time slot $k + 1$ is denoted with $p_h(h'|h)$.

According to the instantaneous utility of the SU in the static jamming game $\mathbf{G}^1$, we can write the long-term utility of the SU in the repeated jamming game $\mathbf{G}^4$ as

$$\mathbf{U}_s(P_s, s) = U_s(P_s, s) + \delta \sum_{s'=0}^{S} \max_{0 \leq P_s' \leq P_s^{max}} \mathbf{U}_s(P_s', s'), \qquad (4.23)$$

where the first term is the instantaneous utility of the SU at time k, and the second term represents the expected utility of the SU in the next time slot.

4.5 Jamming Games with Incomplete Information

The players in the jamming game do not always have complete information of other's actions and system parameters. Thus we model the interactions between a legitimate transmitter and a jammer as a power control Stackelberg game based on prospect theory (PT), in which the leader is the transmitter that chooses its transmit power in advance and the follower is the jammer whose optimal strategy is based on the transmitter's strategy. However, given the underlying detection error and the gossip channel propagation error, the action detection error was introduced in the wireless security game in [15]. It assumes that the smart jammer cannot obtain the actual transmitter's strategy but knows the probability of action to be identified successfully. The decision weighting function developed in [16] can formulate the subjectivity of the both players regarding the opponent's action.

Finite discrete power levels are used by the transmitter and continuous power levels are used by the jammer. The action of the transmitter is denoted by $a_t \in \{p_m\}_{1 \leq m \leq K}$, where p_m is the transmission power with $0 \leq p_m \leq P_t^{max}$. Similarly, the jamming power is given by a_j, $0 \leq a_j \leq P_j^{\text{max}}$. For simplicity, we assume $0 \leq p_m < p_n \leq P_t^{max}$, $\forall 1 \leq m < n \leq K$, where P_x^{max} is maximum power of player x.

The interactions between those players are formulated as a non-zero-sum non-cooperative game $\mathbf{G}^5$. The transmitter takes action p_m first and then the jammer chooses q. Let C_x be the transmission cost per unit power of player x, $x = t, j$. The SINR is considered in the utility of the transmitter, and $u_x(m, n)$ is the utility of player x with $a_t = p_m$ and $a_j = q$, given by

$$u_t(p_m, q) = \frac{h_t p_m}{\sigma + h_j q} - C_t p_m, \tag{4.24}$$

$$u_j(p_m, q) = -\frac{h_t p_m}{\sigma + h_j q} - C_j q, \tag{4.25}$$

where h_t (or h_j) is the channel gain between the transmitter (or jammer) and the receiver, and σ is the noise power.

However, due to the propagation error of the gossip channel and the behavior detection error, the jammer has uncertainty regarding the transmission power of the transmitter. The probability distribution matrix of the action detection process is denoted by $\Phi = [\varepsilon_{m,n}]_{0 \leq m,n \leq K}$, where $\varepsilon_{m,n}$ is the probability that action m is taken as n. For simplicity, we assume $\varepsilon_{m,m} = \varepsilon$, $\forall m$, and $\varepsilon_{m,n} = \frac{1-\varepsilon}{K-1}$, if $m \neq n$, which is known to all the players in the game.

Prospect theory can be used to in user-centric jamming game to evaluate the subjective nature of human decision-making process [17, 18]. Given the action detection error, prospect theory was applied in the Stackelberg anti-jamming game with Perlec's probability weight function [19, 20]. Subjective players tend to under-weigh the high probability outcomes and over-weigh the low probability outcomes. The probability weight function of player x, $x = t, j$, denoted by $w_x(\varepsilon)$, is given by

$$w_x(\varepsilon) = \exp\left(-(-\ln \varepsilon)^{\alpha_x}\right), 0 < \alpha_x \leq 1, \tag{4.26}$$

where α_x is the objective weight of players x and decreases with its subjectivity. The probability weight function is shown in Fig. 4.1. In the special case with $\alpha_x = 1$, $w_x(\varepsilon) = \varepsilon$, indicates that the player is objective.

The expected utility of the transmitter (or jammer) averaged over all the realizations of the opponent's action denoted by U_t^{EUT} (or U_j^{EUT}), is given by

$$U_t^{EUT}\left(a_t = p_m, a_j = q\right) = \sum_{i=1}^{K} \varepsilon_{m,i}\left(\frac{h_t p_i}{\sigma + h_j q} - C_t p_i\right), \tag{4.27}$$

$$U_j^{EUT}\left(a_t = p_m, a_j = q\right) = \sum_{i=1}^{K} \varepsilon_{m,i}\left(-\frac{h_t p_i}{\sigma + h_j q} - C_j q\right). \tag{4.28}$$

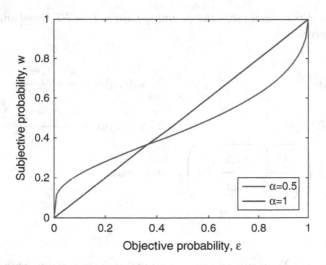

Fig. 4.1 Probability weight function

With unknown detection errors of the opponent's action, a subjective transmitter (or jammer) chooses its transmission power to maximize its prospect theory-based utility, denoted by U_t^{PT} (or U_j^{PT}), which is given by

$$U_t^{PT}\left(a_t = p_m, a_j = q\right) = \sum_{i=1}^{K} w_t(\varepsilon_{m,i})\left(\frac{h_t p_i}{\sigma + h_j q} - C_t p_i\right), \qquad (4.29)$$

$$U_j^{PT}\left(a_t = p_m, a_j = q\right) = \sum_{i=1}^{K} w_j(\varepsilon_{m,i})\left(-\frac{h_t p_i}{\sigma + h_j q} - C_j q\right). \qquad (4.30)$$

The transmitter first chooses its transmit power $p_m \in \{p_i\}_{1 \le i \le K}$, based on the weight probability $w_j(\varepsilon_{m,i})$ of the subjective jammer. The subjective transmitter aims to minimize the utility of the jammer and knows the decision-making process of the jammer.

At the SE of the PT-based anti-jamming game $\mathbf{G}^5$ the subjective smart jammer chooses its jamming power according to the estimated transmit power of the transmitter with detection errors to maximize its PT-based utility U_j^{PT}.

The players choose their power levels to maximize their individual PT-based utilities, thus the SE strategies in the game are given by

$$a_t^{SE} = \arg \max_{a_t \in \{p_m\}_{1 \le m \le K}} U_t^{PT}(a_t, a_j^*), \qquad (4.31)$$

$$a_j^{SE} = \arg \max_{0 \le a_j \le P_j^{\max}} U_j^{PT}(a_t, a_j). \qquad (4.32)$$

Lemma 4.4. *The optimal subjective jamming power of the PT-based anti-jamming game* $\mathbf{G}^5$ *is given by*

$$
a_j^{SE} =
\begin{cases}
0, & \sum_{i=1}^{K} w_j(\varepsilon_{m,i})p_i < \dfrac{\sigma^2 C_j \sum_{i=1}^{K} w_j(\varepsilon_{m,i})}{h_t h_j} \\[4ex]
P_j^{max}, & \sum_{i=1}^{K} w_j(\varepsilon_{m,i})p_i \geq \dfrac{\left(h_j P_j^{max}+\sigma\right)^2 C_j \sum_{i=1}^{K} w_j(\varepsilon_{m,i})}{h_t h_j} \\[4ex]
\dfrac{1}{h_j}\left(\sqrt{\dfrac{h_t h_j \sum_{i=1}^{K} w_j(\varepsilon_{m,i})p_i}{C_j \sum_{i=1}^{K} w_j(\varepsilon_{m,i})}}-\sigma\right), & \text{otherwise.}
\end{cases}
$$

$$(4.33)$$

Proof. See [14].

As shown in (4.33), if subjectively assuming that the transmitter has a low power level compared with a threshold based on its jamming cost, the jammer chooses $a_j^{SE} = 0$, i.e., not to block the ongoing transmission. On the other hand, the jammer subjectively applies its highest jamming power, if the jamming cost is negligible compared with a threshold based on the transmit power of the transmitter.

Lemma 4.5. *The optimal subjective transmission strategy of the PT-based anti-jamming game* $\mathbf{G}^5$ *is given by*

$$
a_t^{SE} =
\begin{cases}
0, & \Gamma_1 \\[2ex]
\underset{a_t \in \{p_m\}_{1 \leq m \leq K}}{\arg\min} \left| \dfrac{\sigma^2 C_j \sum_{i=1}^{K} w_j(\varepsilon_{m,i})}{h_t h_j} - \sum_{i=1}^{K} w_t(\varepsilon_{m,i})p_i \right|, & \Gamma_2 \\[3ex]
\underset{a_t \in \{p_m\}_{1 \leq m \leq K}}{\arg\min} \left| \dfrac{h_t C_j \sum_{i=1}^{K} w_j(\varepsilon_{m,i})}{4 h_j C_t^2} - \sum_{i=1}^{K} w_t(\varepsilon_{m,i})p_i \right|, & \Gamma_3 \\[3ex]
P_t^{max}, & \text{otherwise,}
\end{cases}
$$

$$(4.34)$$

where

$$
\Gamma_1 : \frac{h_t}{\sigma} \leq C_t, \text{ or } \frac{h_t}{2\sigma} \leq C_t < \frac{h_t}{\sigma}, \; C_j < \frac{h_t h_j \sum_{i=1}^{K} w_t(\varepsilon_{1,i})p_i}{\sigma^2 \sum_{i=1}^{K} w_j(\varepsilon_{m,i})},
$$

$$
\text{or } \frac{h_t}{h_j P_j^{max}+\sigma} \leq C_t < \frac{h_t}{2\sigma}, \; C_j < \frac{4 h_j C_t^2 \sum_{i=1}^{K} w_t(\varepsilon_{1,i})p_i}{h_t \sum_{i=1}^{K} w_j(\varepsilon_{m,i})},
$$

$$
\text{or } \frac{h_t}{2\left(h_j P_j^{max}+\sigma\right)} \leq C_t < \frac{h_t}{h_j P_j^{max}+\sigma},
$$

$$\frac{h_t h_j \sum_{i=1}^{K} w_t(\varepsilon_{K,i}) p_i}{\left(h_j P_j^{max} + \sigma\right)^2 \sum_{i=1}^{K} w_j(\varepsilon_{m,i})} < C_j < \frac{4 h_j C_t^2 \sum_{i=1}^{K} w_t(\varepsilon_{1,i}) p_i}{h_t \sum_{i=1}^{K} w_j(\varepsilon_{m,i})},$$

$$or \quad \frac{h_t}{2\left(h_j P_j^{max} + \sigma\right)} \leq C_t < \frac{h_t}{h_j P_j^{max} + \sigma}, \quad U_t^{PT}(a_t = 0) > U_t^{PT}(a_t = P_t^{max}),$$

$$\frac{h_t h_j \sum_{i=1}^{K} w_t(\varepsilon_{1,i}) p_i}{\left(h_j P_j^{max} + \sigma\right)^2 \sum_{i=1}^{K} w_j(\varepsilon_{m,i})} < C_j$$

$$< \min\left(\frac{4 h_j C_t^2 \sum_{i=1}^{K} w_t(\varepsilon_{1,i}) p_i}{h_t \sum_{i=1}^{K} w_j(\varepsilon_{m,i})}, \frac{h_t h_j \sum_{i=1}^{K} w_t(\varepsilon_{K,i}) p_i}{\left(h_j P_j^{max} + \sigma\right)^2 \sum_{i=1}^{K} w_j(\varepsilon_{m,i})}\right);$$

$$\Gamma_2 : \frac{h_t}{2\sigma} \leq C_t < \frac{h_t}{\sigma}, \quad \frac{h_t h_j \sum_{i=1}^{K} w_t(\varepsilon_{1,i}) p_i}{\sigma^2 \sum_{i=1}^{K} w_j(\varepsilon_{m,i})} < C_j < \frac{h_t h_j \sum_{i=1}^{K} w_t(\varepsilon_{K,i}) p_i}{\sigma^2 \sum_{i=1}^{K} w_j(\varepsilon_{m,i})};$$

$$\Gamma_3 : \frac{h_t}{\sigma + h_j P_j^{max}} \leq C_t < \frac{h_t}{2\sigma},$$

$$\frac{4 h_j C_t^2 \sum_{i=1}^{K} w_t(\varepsilon_{1,i}) p_i}{h_t \sum_{i=1}^{K} w_j(\varepsilon_{m,i})} < C_j < \frac{4 h_j C_t^2 \sum_{i=1}^{K} w_t(\varepsilon_{K,i}) p_i}{h_t \sum_{i=1}^{K} w_j(\varepsilon_{m,i})},$$

$$or \quad \frac{h_t}{2\left(h_j P_j^{max} + \sigma\right)} \leq C_t < \frac{h_t}{\sigma + h_j P_j^{max}},$$

$$\frac{4 h_j C_t^2 \sum_{i=1}^{K} w_t(\varepsilon_{K,i}) p_i}{h_t \sum_{i=1}^{K} w_j(\varepsilon_{m,i})} > C_j$$

$$> \max\left(\frac{4 h_j C_t^2 \sum_{i=1}^{K} w_t(\varepsilon_{1,i}) p_i}{h_t \sum_{i=1}^{K} w_j(\varepsilon_{m,i})}, \frac{h_t h_j \sum_{i=1}^{K} w_t(\varepsilon_{K,i}) p_i}{\left(h_j P_j^{max} + \sigma\right)^2 \sum_{i=1}^{K} w_j(\varepsilon_{m,i})}\right),$$

$$or \quad \frac{h_t}{2\left(h_j P_j^{max} + \sigma\right)} \leq C_t < \frac{h_t}{\sigma + h_j P_j^{max}},$$

$$U_t^{PT}(a_t = P_t^{max}) < U_t^{PT}\left(\sum_{i=1}^{K} w_t(\varepsilon_{m,i}) p_i = \frac{h_t C_j \sum_{i=1}^{K} w_j(\varepsilon_{m,i})}{4 h_j C_t^2}\right),$$

$$\frac{4 h_j C_t^2 \sum_{i=1}^{K} w_t(\varepsilon_{1,i}) p_i}{h_t \sum_{i=1}^{K} w_j(\varepsilon_{m,i})} < C_j < \frac{h_t h_j \sum_{i=1}^{K} w_t(\varepsilon_{K,i}) p_i}{\left(h_j P_j^{max} + \sigma\right)^2 \sum_{i=1}^{K} w_j(\varepsilon_{m,i})}. \tag{4.35}$$

Proof. See [14].

The SE of the PT-based anti-jamming game $\mathbf{G}^5$ is given by (a_t^{SE}, a_j^{SE}) in Lemmas 4.4 and 4.5. The smart jammer allocates its jamming power according to the jamming cost, channel states and the subjectivity regarding the detection error of the transmission power. The subjective transmitter adjusts its strategy according to the transmission cost and the jammer's reward.

To compare with the Stackelberg equilibria, the Nash equilibria of the PT-based anti-jamming game is provided. Let (a_t^{NE}, a_j^{NE}) denote the NE of the static game in which both players make decision at the same time. At a NE of the game, given the power level of the opponent, no player can improve its utility by changing its power unilaterally.

$$U_t^{PT}(a_t^{NE}, a_j^{NE}) \geq U_t^{PT}(a_t, a_j^{NE}), \tag{4.36}$$

$$U_j^{PT}(a_t^{NE}, a_j^{NE}) \geq U_j^{PT}(a_t^{NE}, a_j). \tag{4.37}$$

Lemma 4.6. *The NE of the PT-based anti-jamming game* $\mathbf{G}^5$ *is given by*

$$(a_t^{NE}, a_j^{NE}) = \begin{cases} \left(P_t^{max}, P_j^{max} \right), & I_1 \\ \left(P_t^{max}, 0 \right), & I_2 \\ \left(0, P_j^{max} \right), & I_3 \\ (0,0), & I_4 \\ \left(P_t^{max}, \frac{1}{h_j} \left(\sqrt{\frac{h_j h_t \sum_{i=1}^{K} w_j(\varepsilon_{K,i}) p_i}{C_j \sum_{i=1}^{K} w_j(\varepsilon_{K,i})}} - \sigma \right) \right), & I_5 \\ \left(0, \frac{1}{h_j} \left(\sqrt{\frac{h_j h_t \sum_{i=1}^{K} w_j(\varepsilon_{K,i}) p_i}{C_j \sum_{i=1}^{K} w_j(\varepsilon_{K,i})}} - \sigma \right) \right), & I_6 \\ \left(a_t^{opt}, \frac{1}{h_j} \left(\frac{h_t}{C_t} - \sigma \right) \right), & otherwise, \end{cases} \tag{4.38}$$

where

$$a_t^{opt} = \underset{a_t \in \{p_m\}_{1 \leq m \leq K}}{\arg\min} \left| \frac{h_t C_j \sum_{i=1}^{K} w_i(\varepsilon_{m,i})}{h_j C_t^2} - \sum_{i=1}^{K} w_t(\varepsilon_{m,i}) p_i \right|,$$

$$I_1 : C_t < \frac{h_t}{\sigma + h_j P_j^{max}}, \quad C_j < \frac{h_j h_t \sum_{i=1}^{K} w_j(\varepsilon_{K,i}) p_i}{\left(\sigma + h_j P_j^{max} \right)^2 \sum_{i=1}^{K} w_j(\varepsilon_{K,i})};$$

$$I_2 : C_t < \frac{h_t}{\sigma + h_j P_j^{max}}, \quad C_j > \frac{h_j h_t \sum_{i=1}^{K} w_j(\varepsilon_{K,i}) p_i}{\sigma^2 \sum_{i=1}^{K} w_j(\varepsilon_{K,i})},$$

$$or \quad \frac{h_t}{\sigma + h_j P_j^{max}} \leq C_t \leq \min\left(\frac{h_t}{\sigma}, \sqrt{\frac{h_t C_j \sum_{i=1}^{K} w_j(\varepsilon_{m,i})}{h_j \sum_{i=1}^{K} w_j(\varepsilon_{K,i}) p_i}}\right);$$

$$I_3 : C_t > \frac{h_t}{\sigma}, \quad C_j < \frac{h_j h_t \sum_{i=1}^{K} w_j(\varepsilon_{1,i}) p_i}{\left(\sigma + h_j P_j^{max}\right)^2 \sum_{i=1}^{K} w_j(\varepsilon_{1,i})};$$

$$I_4 : C_t > \frac{h_t}{\sigma}, \quad C_j > \frac{h_j h_t \sum_{i=1}^{K} w_j(\varepsilon_{1,i}) p_i}{\sigma^2 \sum_{i=1}^{K} w_j(\varepsilon_{1,i})};$$

$$I_5 : C_t < \frac{h_t}{\sigma + h_j P_j^{max}}, \quad \frac{h_j h_t \sum_{i=1}^{K} w_j(\varepsilon_{K,i}) p_i}{\left(\sigma + h_j P_j^{max}\right)^2 \sum_{i=1}^{K} w_j(\varepsilon_{K,i})} \leq C_j \leq \frac{h_j h_t \sum_{i=1}^{K} w_j(\varepsilon_{K,i}) p_i}{\sigma^2 \sum_{i=1}^{K} w_j(\varepsilon_{K,i})};$$

$$I_6 : C_t > \frac{h_t}{\sigma}, \quad \frac{h_j h_t \sum_{i=1}^{K} w_j(\varepsilon_{1,i}) p_i}{\left(\sigma + h_j P_j^{max}\right)^2 \sum_{i=1}^{K} w_j(\varepsilon_{1,i})} \leq C_j \leq \frac{h_j h_t \sum_{i=1}^{K} w_j(\varepsilon_{1,i}) p_i}{\sigma^2 \sum_{i=1}^{K} w_j(\varepsilon_{1,i})}.$$

Proof. See [14].

The average SINR and the utilities of the players in the PT-based jamming game depend on the jammer's objective weight α_j, as shown in Eqs. (4.24) and (4.25). Figure 4.2a presents the performance of the PT-based anti-jamming game $\mathbf{G}^5$ with $\alpha_s = \alpha_j$, $C_t = 0.4$, $C_j = 0.33$, $h_t = h_j = 0.8$ and $\sigma = 0.5$, showing that the performance of the SU at the SE and NE of the game decreases with jammer's objectivity (α_j). For instance, the SINR decreases from 0.65 to 0.42 at the SE and from 0.8 to 0.62 at the NE, as α_j changes from 0.5 to 1, with correct detection probability $\varepsilon = 0.7$. The reason is that a subjective jammer is less likely to block the transmission of the SU as the jammer tends to overweigh its loss due to useless jamming. Meanwhile, for a fixed α_j, a higher probability of the action identified successfully (ε) results in a smaller SINR. This is because a subjective jammer with higher probability ε has more clear understanding on the action detection process. In addition, the smart jammer can learn the ongoing transmission flexibly before making decision.

As shown in Fig. 4.2b, the utility of the SU deceases with α_j, while the performance of the jammer increases with α_j, because both subjective players are less likely to send signals to avoid the highly unfavorable loss due to the action detection error. One the other hand, the jammer with the SE strategy is more intelligent than that with the NE strategy. Consequently, the jammer's utility at the SE is higher than that at the NE, while the transmitter's utility is the opposite.

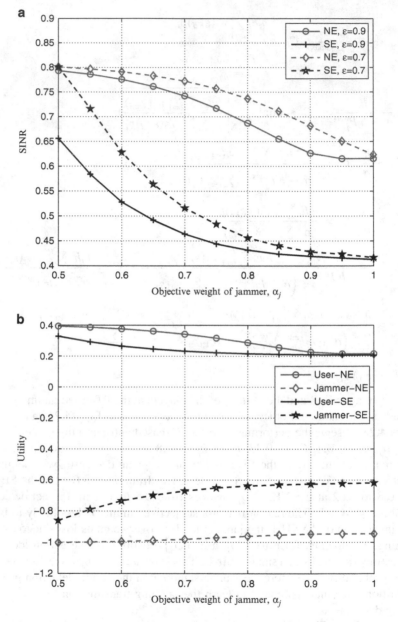

Fig. 4.2 Performance of the SE and NE of the PT-based anti-jamming game $\mathbf{G}^5$ with correct detection probability $\varepsilon = 0.9$. (**a**) SINR, (**b**) Utility

4.6 Summary

We have investigated the Nash equilibrium and Stackelberg equilibrium of static jamming games. Nodes cooperation in the anti-jamming communications has been analyzed with a cooperative anti-jamming game. We have formulated the repeated interactions between a secondary user and a jammer in multiple time slots as a dynamic jamming game. Jamming game with incomplete information has also been analyzed for the case in which secondary users do not know the accurate channel parameters.

References

1. Wang, Q., Xu, P., Ren, K., Li, X.: Towards optimal adaptive UFH-based anti-jamming wireless communication. IEEE J. Sel. Areas Commun. **30**(1), 16–30 (2012)
2. Chen, C., Song, M., Xin, C., Backens, J.: A game-theoretical anti-jamming scheme for cognitive radio networks. IEEE Netw. **27**(3), 22–27 (2013)
3. Wu, Y., Wang, B., Liu, K.J.R., Clancy, T.: Anti-jamming games in multi-channel cognitive radio networks. IEEE J. Sel. Areas Commun. **30**(1), 4–15 (2012)
4. Zhu, Q., Li, H., Han, Z., Basar, T.: A stochastic game model for jamming in multi-channel cognitive radio systems. In: Proceedings of IEEE International Conference on Computer Communication, pp. 1–6, May 2010
5. Sagduyu, Y.E., Ephremides, A.: A game-theoretic analysis of denial of service attacks in wireless random access. Wirel. Netw. **15**(5), 651–666 (2009)
6. Altman, E., Avrachenkov, K., Garnaev, A.: Jamming in wireless networks under uncertainty. Mob. Netw. Appl. **16**(2), 246–254 (2011)
7. Nguyen, K.C., Alpcan, T., Başar, T.: Security games with incomplete information. In: Proceedings of IEEE International Conference on Computer Communication, pp. 1–6, June 2009
8. Gupta, A., Nayyar, A., Langbort, C., Basar, T.: A dynamic transmitter-jammer game with asymmetric information. In: Proceedings of IEEE Annual Conference on Decision and Control, pp. 6477–6482 (2012)
9. Yang, D., Xue, G., Zhang, J., Richa, A., Fang, X.: Coping with a smart jammer in wireless networks: a Stackelberg game approach. IEEE Trans. Wirel Commun. **12**(8), 4038–4047 (2013)
10. Xiao, L., Chen, T., Liu, J., Dai, H.: Anti-jamming transmission Stackelberg game with observation errors. IEEE Trans. Commun. Lett. **19**(6), 949–952 (2015)
11. Li, Y., Xiao, L., Liu, J., Tang, Y.: Power control Stackelberg game in cooperative anti-jamming communications. In: Proceedings of IEEE Game Theory for Networks (GAMENETS), pp. 1–6 (2014)
12. Khalil, K., Ekici, E.: Multiple access game with a cognitive jammer. In: Proceedings of Asilomar Conference on Signals, Systems and Computers (ASILOMAR), pp. 1383–1387, Nov 2012
13. Zhang, G., Li, P., Zhou, D., Yang, K., Ding, E.: Optimal power control for wireless cooperative relay networks: a cooperative game theoretic approach. Int. J. Commun. Syst. **26**(11), 1395–1408 (2013)
14. Liu, J., Xiao, L., Li, Y., Huang, L.: User-centric analysis on jamming game with action detection error. In: Proceedings of IEEE Game Theory for Networks (GAMENETS), pp. 1–6 (2014)

15. Xiao, L., Chen, Y., Lin, W.S., Liu, K.: Indirect reciprocity security game for large-scale wireless networks. IEEE Trans. Inf. Forensics Secur. **7**(4), 1368–1380 (2012)
16. Prelec, D.: The probability weighting function. Econometrica **66**(3), 497–527 (1998)
17. Li, T., Mandayam, N.B.: Prospects in a wireless random access game. In: Proceedings of IEEE Information Sciences and Systems (CISS), pp. 1–6 (2012)
18. Xiao, L., Liu, J., Li, Y., Mandayam, N.B., Poor, H.V.: Prospect theoretic analysis of anti-jamming communications in cognitive radio networks. In: Proceedings of IEEE Global Communications Conference (GLOBECOM), pp. 746–751 (2014)
19. Kahneman, D., Tversky, A.: Prospect theory: an analysis of decision under risk. Econometrica J. Econ. Soc. **47**, 263–291 (1979)
20. Tversky, A., Kahneman, D.: Advances in prospect theory: cumulative representation of uncertainty. J. Risk Uncertain. **5**(4), 297–323 (1992)

Chapter 5
Game Theoretic Stimulation Mechanisms

5.1 Introduction

With the development of cognitive radio networks, a large number of secondary users autonomously and independently use opportunistic radio access in large-scale networks with limited spectrum resources, each aiming to maximize its individual utilities. A selfish secondary user can obtain illegal advantages by blocking the ongoing transmissions of the users that he/she dislikes. Due to the dynamic network topology, the opportunistic radio connection, weak unforgeability for nodes, and lack of centralized monitoring, a variety of attacks ranging from passive eavesdropping to active interfering (e.g., malicious jamming) can be launched in cognitive radio networks.

Game theory has shown its strength to formulate the interactions among users in cognitive radio networks [1]. In the games, players interact with each other with objectives to maximize their individual interests, especially in resource allocation and service providing. Particularly, in noncooperative games, each player aims to maximize its utility without considering the performance of other players, which sometimes is detrimental to the network. To improve the anti-jamming performance of cognitive radio network, several game theoretic mechanisms have been applied.

First, pricing mechanisms reviewed in Sect. 5.2 can be used to manage bandwidth, spectrum access and transmit power allocation in wireless networks [2]. In Sect. 5.3, we consider auctions, in which buyers submit bids for their interested network commodities and the seller makes a deal at the end of the auction [3]. Finally, in Sect. 5.4, we present trust and reputation systems. Direct reciprocity is applied to describe the repeated interactions between two players and defend against attacks in small size networks [4], while indirect reciprocity is used to address jamming in large-scale networks with node mobility.

L. Xiao, *Anti-Jamming Transmissions in Cognitive Radio Networks*,
SpringerBriefs in Electrical and Computer Engineering,
DOI 10.1007/978-3-319-24292-7_5

5.2   Pricing Mechanisms

Pricing involves two types of participants, including sellers and buyers. A seller designs efficient pricing strategy to attract customers to achieve profits. Pricing in economy depends on the manufacturing cost, market, competition, and quality of product, etc [5]. The networks with pricing usually involve service providers and subscribers, depending on the market structure of the network service, radio regulatory environment, cost of relevant technologies and network capacity [6]. As shown in Table 5.1, flat rate pricing is mainly used to set a fixed price independently from environment such as network conditions, while parameter-based pricing can charge services flexibly and accurately [2].

Pricing is effective for radio resource management in wireless networks because of its ability to guide user behavior toward a more efficient operating point. The interactions between a seller and a buyer can be formulated as a game, in which the seller sets an optimal pricing strategy for its scarce resources and the buyer determines whether to take it according to the expected profit.

A linear pricing-based wireless system proposed in [12] improves the energy efficiency and reduces interference under spectrum constraints, in which each node has to pay an amount proportional to its transmit power. Another linear pricing scheme based on the signal-to-interference ratio designed in [13] achieves Pareto-efficiency in power control. As spectrum is under-utilized in wireless community, licensed spectrum of primary users can be sold to secondary users and the latter is charged according to the spectrum usage. In [14], the primary service provider (e.g., base station) enhances its revenue via charging secondary users on their transmit power levels. According to the importance of the resources, the convex pricing scheme presented in [15] allocates transmit power for multi-service in the uplink, leading to a unique Pareto optimal Nash equilibrium. Stackelberg and bargaining games investigated in [10] achieve better fairness, in which users buy power to improve their transmission rates from the relay that sets price to maximize its revenue. In the power control games based on pricing, players attempt to maximize their individual revenues and achieve fairness by charging the transmit power and power-related constraints such as SNR and SINR.

Pricing mechanisms can also stimulate user collaboration in spectrum utilization [16, 17], wireless channel access [18, 19], and routing [20–22]. For instance, the dynamic spectrum sharing scheme investigated in [16] derives the optimal

Table 5.1 Taxonomy of pricing schemes

Class	Characteristic	Examples
Flat rate pricing	Fixed price	Packet pricing, time-based pricing [7]
Parameter-based pricing	Flexible and accurate price	Static pricing: adaption [8], cumulus [9]
		Dynamic pricing: game theory [10], probability [11], auction [3]

pricing for CRNs, in which multiple primary users compete for spectrum access opportunities to secondary users and collusions among primary users can be avoided in repeated interactions. A cooperative spectrum sharing between a primary user and multiple secondary users in CRN was formulated as a Nash bargaining problem in [17]. In a twofold pricing scheme in [18], a relay system based on the interference pricing was designed to encourage nodes to relay packets. In multi-hop relay networks, a pricing game formulated in [20] provides incentives to forward traffic and yeilds the optimal network routing. Based on the idea of "pay for cooperation", a pricing-based relay framework was designed in [21], in which each flow (i.e., a source-destination pair) offers a payment for successfully received packets and nodes that relay a packet share the payment. A pricing-based routing in CRN proposed in [22], applies the log-based pricing based on the SNR of the buyer.

Unlike selfish users, malicious users do not request for network services and thus throw more serious threats to wireless networks. In [23], a two-tier jamming based on a given pricing parameter for covert timing networks deploys camouflaging resources to address jamming attacks. The congestion pricing in [24] mitigates DoS attack, in which users pay the allocated bandwidth against DoS attackers.

5.3 Auction Mechanisms

As a market clearing mechanism, auction theory describes a problem that a seller attempts to obtain the highest expected revenue through selling an object with unknown value to the buyer. Due to the rareness or uniqueness of the object such as spectrum in wireless networks, scarce resources are allocated to the individual who bids at the highest price. In contrast to the fixed price and price negotiation mechanisms, auctions are more flexible and consume less time. In general, a seller can sell the object fast with best return.

A seller possesses sales resources, such as time slots, usage right of spectrum and bandwidth in wireless communication auctions. A buyer intends to buy resources from the seller by bidding price, e.g., a buyer may aim to buy radio resources for transmission tasks in wireless networks. Both buyers and sellers are players in auction games. An auctioneer acts as an intermediary to conduct actions and generally is a seller itself. A merchandise in an auction has its value and is traded between a seller and a buyer. A seller or a buyer has reserved monetary evaluation of a merchandise, which varies among players. A seller submits an ask, i.e., the price for a merchandise, while a buyer can submit a bid, i.e., the bidding price for the desired merchandise. Finally, an auctioneer determines the hammer price [3].

In Table 5.2, auctions can have either a single object or multiple objects to trade. Price trend classifies auctions into ascending price auctions starting at a low price and descending price auctions starting at a high price. Based on the publicity of bids, we have open auctions in which bids are public and sealed-bid auctions that

Table 5.2 Taxonomy of auctions

Distinct criteria	Class	Examples
Number of object	Single object	English [25], Dutch [25], first-price [26]
	Multiple objects	Combinatorial auctions [27], sequential actions [28]
Price trend	Ascending price auctions	Public bids starting at a low price
	Descending price auctions	Public bids starting at a high price
Publicity of bid	Open auctions	All bids are publicly observable
	Sealed-bid auctions	All bids are not publicly observable
Objectivity	Forward auctions	Buyers bid for commodities from seller(s)
	Reverse auctions	Sellers compete for buyer(s) patronizing

keep bids non-public. In addition, there are forward auctions in which buyers bid for commodities from sellers and reverse auctions in which sellers compete for buyers patronizing [25].

As auctions depend on the decisions of sellers and buyers, sale object, knowledge of bids and special rules to determine the winning bid, game theory can formulate the interactions in auctions [25]. In auction games, both the sellers and buyers choose their own auction bidding strategies according to the knowledge of the current auction mechanism and other participants. How to design efficient and beneficial auction mechanisms (i.e., auction rules) based on game theory has attracted extensive attentions in wireless communications [27–34].

A sequential second price auction-based bandwidth and power allocation was presented in [28], in which resources are partitioned and allocated sequentially in many rounds and the auctioneer allocates resources to the agent that submits the largest bid and charges the agent the second largest bid. As an extensive form game with a balance game tree, the auction game has an efficient dominant strategy equilibrium, significantly saving the worst-case efficiency loss. With small computation and information exchange among the resource owner and agents, sequential auction can be applied to scenarios where agents join and leave the auctions at any time in [29, 30]. Although sequential actions do not always achieve an efficient allocation, researchers have analyzed how to reduce the worst-case efficiency loss [31].

With the explosive growth of the number of network users, traditional auction frameworks are vulnerable to security risks, because selfish users maximize their individual interests and sometimes maliciously cheat in auctions. For instance, selfish users can occupy more idle channels and reduce the access opportunity of legitimate users, resulting lower revenue of the entire system. Therefore, secure auctions in current vulnerable wireless networks have attracted the attentions of researchers [27, 32–40].

In [32], a belief-assisted double auction mechanism was proposed to achieve efficient dynamic spectrum allocation, and collusion-resistant strategies combat possible user collusive behavior using optimal reserve prices. In [33], a multi-winner spectrum auction based on interference-limited resources improves the spectrum efficiency, which is robust against loser collusions, sublease collusions and kick-out collusions (i.e., a clique of secondary users cheat together to pursuit of higher profits by distorting valuations of spectrum resources) and scalable for a large-scale network.

The auction-based spectrum allocation proposed in [35] for multimedia streaming in CRNs applies single object pay-as-bid ascending clock auction and alternative ascending clock auction to prevent cheating and enforce selfish secondary users to report their true demands at every clock. Other auction-based approaches such as the Vickrey Clarke Groves (VCG) mechanism in [37] provide assurance of truthfulness to address lies from secondary users.

Sellers in auctions sometimes overcharge the resource buyers by deceiving, e.g., the back-room dealing by the insincere auctioneer in the spectrum auction of CRN which can be addressed via homomorphic encryption in [27] and paillier cryptosystem in [34]. A physical-layer security mechanism against malicious eavesdropper in [38] exploits friendly jamming to interfere adversary, in which the jammer is the auctioneer and the source node is the bidder in the power allocation. In order to motivate the participation of cooperative jammers, in [39], legitimate parties give the friendly jammers an opportunity to use the fraction of their spectrum as a compensation for the invested power. In the presence of malicious jamming attacks and selfish secondary users participating in the auction for the unjammed spectrum, a zero-sum stochastic game and randomized two-level auctions were proposed in [40], in which the auctioneer allocates vacant channels in the first auction and assigns the remaining channels in the second auction.

5.4 Reputation and Trust

The reputation in [41] is a value that is allocated to each entity according to its action history and determines its probability to receive the network services in the future. More specifically, a higher reputation means more weight in the future decision-making and the transmitter with a higher reputation value is more likely to obtain node cooperation in the networks [1]. Trust is subjective measure of belief that one entity as the subject assesses the probability that another entity as performs a favourable action before such action [42]. More specifically, the subject measures the trust value by evaluating the uncertainty that the agent's action performed in the subject's point of view. Reputation is one of ways to create a trust between participants.

In distributed networks, the collaborations among distributed entities have a significant impact on the network performance. Theory of trust and reputation-based trust mechanisms can be used to improve communication efficiency and security. Trust has been measured based on common properties such as integrity, ability and benevolence in [43] and such measurements have been used in [44] to secure collaboration in uncertain environment. Trust mechanism in communication networks has been applied in wireless ad hoc [45], sensor networks [46], and peer-to-peer networks [47]. For instance, theoretic definitions of terms related to trust and reputation in CRNs were presented in [48] and a brief discussion on the trust model in CRNs was provided in [49]. In [50], a trust aware hybrid spectrum sensing scheme can detect misbehaving secondary users.

Important and critical security threats against distributed networks were surveyed in [51] and a new trust and reputation mechanism was developed to defend these threats. In most security scenarios, entities in networks have the ability to evaluate trust to motivate incentive for good behaviors and record historical trust values for the past behaviors of entities. Hence, the subjects can make decisions, i.e., appropriate actions, by predicting the future actions of agents based on corresponding trust evaluation and avoid cooperating with less trustworthy entities. Therefore, the trust evaluation gives insights into addressing attacks launched by adversaries who have low trust in networks [1].

Due to the similarity between wireless networks and social networks, such as a prevalent reputation and social norm system [52], reciprocity principle can stimulate cooperation among entities in CRNs. In order to discriminate selfish entities in the cooperation, reciprocity-based nodes establish trust systems, which can be extended to reputation systems. Direct reciprocity principle requires entities to record their bilateral interactions in the trust systems, while indirect reciprocity expects entities to track interactions between other network participants [53].

5.4.1   Direct Reciprocity

The main idea of direct reciprocity principle is "being nice to others who are nice to you" [4], as shown in Fig. 5.1. More specifically, direct reciprocity stimulates altruism among two players based on the repeated encounters between the same individuals. In addition, both individuals must be able to provide help. Each node chooses its actions according to the interaction history with its opponents, and is more likely to decline the requests from those who have ever attacked it before. Direct reciprocity is a non-distributed mechanism that one player makes decision according to the belief based on previous interactions with the opponent. Based on the history, both players tend to promote the mutual cooperation and the current reputation of the opponent is described as the belief about the new cooperative

Fig. 5.1 Illustration of direct reciprocity principle, in which Alice helps Bob, because the latter has helped Alice

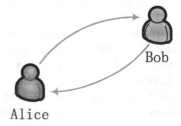

interaction. The repeated Prisoner's Dilemma is the standard model of direct reciprocity. In each round of the repeated game, a player chooses to cooperate or not based on what the opponent has done [54].

With the trust modeling and evaluation method proposed in [55], the reciprocity mechanism has become a powerful tool to improve security and stimulate cooperation in wireless networks, e.g., to address the vulnerabilities in trust establishment methods [56], insider attacks [57] and Byzantine attacks [58].

Direct reciprocity is a powerful mechanism in the evolution of cooperation. In mobile ad hoc networks, direct reciprocity has been used in packet relay with the idea "I relay your packet and you will in the future relay mine" [59]. The interactions between subject entities and agents that takes risky or cooperative actions were formulated as a social game in [60], in which the agent with risky behavior is punished by the subject entity with a tit-for-tat strategy.

However, as interactions between the players can be asymmetric and fleeting, direct reciprocity principle sometimes fails to stimulate altruism, especially in large-scale networks with terminal mobility. In this case, most secondary users have a small chance to meet their opponents frequently, resulting in limited or outdated knowledge on the interaction histories of their current opponents. Therefore, attackers rarely meet their victims again and are punished according to direct reciprocity principle. Consequently, the direct reciprocity in small-scale networks without mobility can effectively address jamming attacks.

5.4.2 Indirect Reciprocity

Indirect reciprocity was first developed in social science and evolutionary biology, with the main idea "I help you and somebody else helps me" [4], as shown in Fig. 5.2. More specifically, indirect reciprocity stimulates altruism among unrelated individuals without repeated interaction and entities obtain the information regarding potential cooperators from other entities in networks. The mechanism is operated by exchanging the reputation information among participating entities. At the end of each cooperative interaction, the players broadcast the new reputation information of their opponents to the rest of the network. Therefore, each player gains a global view of the network. Indirect reciprocity is robust and guarantees more stable and socially efficient cooperation compared with direct reciprocity [54].

The mechanism in [61] stimulates cooperations in large scale networks. Motivated by ubiquitous WiFi networks distributed in densely populated cities, indirect reciprocity was used in [62] to form a club, in which club members offer other members free WiFi access and are treated equally if they need WiFi. Signed digital receipts were used as an input of such indirect reciprocity to secure sharing. In mobile ad hoc networks, the indirect reciprocity system proposed in [63] stimulates nodes to participate in packet forwarding. A threshold of benefit-to-cost ratio

Fig. 5.2 Illustration of
indirect reciprocity principle,
in which Alice helps Bob,
because the latter has helped
Kate

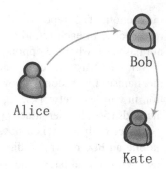

in [64] guarantees the convergence of cooperation. In a multiuser cooperative communication network, an indirect reciprocity game proposed in [65] provides incentives for selfish users to cooperate.

To provide a more reasonable and realistic cooperative scenario in next-generation wireless networks, a reputation auction framework based on indirect reciprocity in [66] fosters node cooperations and reduces the total energy consumption. In cognitive radio networks, the cooperation stimulation between primary users and secondary users was modeled as an indirect reciprocity game in [67], in which SUs were motivated to help relay the PUs' information with reputations based on the vacant spectrum.

Indirect reciprocity improves the Sybil-resistance for the accounting of peer contributions in peer-to-peer networks [68]. Each Sybil attack that claims to have multiple identities other than itself to obtain illegal advantages in various aspects such as receiving more network services or more weights in the network voting is punished by the other nodes.

In the indirect reciprocity security game in [69], a wide range of attacks can be addressed in wireless networks, including jamming, spoofing, Sybil, collusion attacks, relay-related attacks such as the packet dropping attacks. The reputation propagation mechanism allows attackers to be recognized and punished by a much larger network, compared with the direct reciprocity system. Consequently, this system can provide a stronger security protection, especially for the large-scale wireless networks with node mobility.

5.5 Summary

In this chapter, game theoretic incentive mechanisms in communication networks were briefly reviewed. Pricing schemes have been used in the resources allocation, suppression of selfish attacks, node cooperations and anti-jamming transmissions. Auction mechanisms can efficiently allocate scarce resource in networks, where the interactions between the sellers and buyers are formulated as auction games. Finally, we have reviewed trust and reputation mechanisms in wireless communication and network security. Direct reciprocity principle can effectively protect small-scale networks, while indirect reciprocity can improve the attack-resistance in large-scale wireless networks.

References

1. Liu, K.J.R., Wang, B.: Cognitive Radio Networking and Security: A Game-Theoretic View. Cambridge University Press, Cambridge (2010)
2. Gizelis, C.A., Vergados, D.D.: A survey of pricing schemes in wireless networks. IEEE Commun. Surv. Tutorials 13(1), 126–145 (2011)
3. Zhang, Y., Lee, C., Niyato, D., Wang, P.: Auction approaches for resource allocation in wireless systems: a survey. IEEE Commun. Surv. Tutorials 15(3), 1020–1041 (2013)
4. Nowak, M.A., Sigmund, K.: Evolution of indirect reciprocity. Nature 437(7063), 1291–1298 (2005)
5. Kreps, D.M.: A Course in Microeconomic Theory. Princeton University Press, Princeton (1990)
6. Cocchi, R., Shenker, S., Estrin, D., Zhang, L.: Pricing in computer networks: motivation, formulation, and example. IEEE/ACM Trans. Networking 1(6), 614–627 (1993)
7. Shenker, S., Clark, D., Estrin, D., Herzog, S.: Pricing in computer networks: reshaping the research agenda. ACM SIGCOMM Comput. Commun. Rev. 26(2), 19–43 (1996)
8. Marbach, P.: Analysis of a static pricing scheme for priority services. IEEE/ACM Trans. Networking 12(2), 312–325 (2004)
9. Hayel, Y., Tuffin, B.: A mathematical analysis of the cumulus pricing scheme. Comput. Netw. 47(6), 907–921 (2005)
10. Cao, Q., Zhao, H.V., Jing, Y.: Power allocation and pricing in multiuser relay networks using Stackelberg and bargaining games. IEEE Trans. Veh. Technol. 61(7), 3177–3190 (2012)
11. Allen, S.M., Whitaker, R.M., Hurley, S.: Personalised subscription pricing for optimised wireless mesh network deployment. Comput. Netw. 52(11), 2172–2188 (2008)
12. Zhijiat, C., Rami, M., Eduard, J.: Pricing in noncooperative interference channels for improved energy efficiency. EURASIP J. Wireless Commun. Netw. 2010(8), 1–12 (2010)
13. Rasti, M., Sharafat, A.R., Seyfe, B.: Pareto-efficient and goal-driven power control in wireless networks: a game-theoretic approach with a novel pricing scheme. IEEE/ACM Trans. Networking 17(2), 556–569 (2009)
14. Yu, H., Gao, L., Li, Z., Wang, X., Hossain, E.: Pricing for uplink power control in cognitive radio networks. IEEE Trans. Veh. Technol. 59(4), 1769–1778 (2010)
15. Tsiropoulou, E.E., Katsinis, G.K., Papavassiliou, S.: Distributed uplink power control in multiservice wireless networks via a game theoretic approach with convex pricing. IEEE Trans. Parallel Distrib. Syst. 23(1), 61–68 (2012)
16. Niyato, D., Hossain, E: Competitive pricing for spectrum sharing in cognitive radio networks: dynamic game, inefficiency of Nash equilibrium, and collusion. IEEE J. Sel. Areas Commun. 26(1), 192–202 (2008)
17. Wu, Y., Song, W.Z.: Cooperative resource sharing and pricing for proactive dynamic spectrum access via Nash bargaining solution. IEEE Trans. Parallel Distrib. Syst. 25(11), 2804–2817 (2014)
18. Mohsenian-Rad, A.-H., Wong, V.W., Leung, V.C.: Two-fold pricing to guarantee individual profits and maximum social welfare in multi-hop wireless access networks. IEEE Trans. Wireless Commun. 8(8), 4110–4121 (2009)
19. Sarikaya, Y., Alpcan, T., Ercetin, O.: Dynamic pricing and queue stability in wireless random access games. IEEE J. Sel. Top. Sign. Proces. 6(2), 140–150 (2012)

20. Xi, Y., Yeh, E.M.: Pricing, competition, and routing in relay networks. In: Proceedings of Annual Allerton Conference on Communication Control, and Computing, pp. 507–514 (2009)
21. Chen, L., Libman, L., Leneutre, J.: Conflicts and incentives in wireless cooperative relaying: a distributed market pricing framework. IEEE Trans. Parallel Distrib. Syst. **22**(5), 758–772 (2011)
22. Khairullah, E.F., Chatterjee, M., Kwiat, K.: Pricing-based routing in cognitive radio networks. In: Proceedings of IEEE Global Communications Conference (GLOBECOM), pp. 908–912 (2013)
23. Anand, S., Sengupta, S., Chandramouli, R.: An attack-defense game theoretic analysis of multi-band wireless covert timing networks. In: Proceedings of IEEE International Conference on Computer and Communication (INFOCOM), pp. 1–9 (2010)
24. Vulimiri, A., Agha, G.A., Godfrey, P.B., Lakshminarayanan, K.: How well can congestion pricing neutralize denial of service attacks? In: ACM SIGMETRICS Performance Evaluation Review, pp. 137–150 (2012)
25. Krishna, V.: Auction Theory. Academic, New York (2009)
26. Harrison, G.W.: Theory and Misbehavior of First-Price Auctions. Am. Econ. Rev. **79**(4), 749–762 (1989)
27. Pan, M., Li, H., Li, P., Fang, Y.: Dealing with the untrustworthy auctioneer in combinatorial spectrum auctions. In: Proceedings of IEEE Global Communications Conference, pp. 1–5 (2011)
28. Bae, J., Beigman, E., Berry, R.A., Honig, M.L., Vohra, R.: Sequential bandwidth and power auctions for distributed spectrum sharing. IEEE J. Sel. Areas Commun. **26**(7), 1193–1203 (2008)
29. Fu, F., Kozat, U.C.: Wireless network virtualization as a sequential auction game. In: Proceedings of IEEE International Conference Computer and Communications (INFOCOM), pp. 1–9 (2010)
30. Xu, C., Song, L., Han, Z., Zhao, Q., Wang, X., Jiao, B.: Interference-aware resource allocation for device-to-device communications as an underlay using sequential second price auction. In: Proceedings of IEEE International Conference Communications (ICC), pp. 445–449 (2012)
31. Bae, J., Beigman, E., Berry, R., Honig, M.L., Vohra, R.: On the efficiency of sequential auctions for spectrum sharing. In: Proceedings of International Conference Game Theory for Networks, pp. 199–205 (2009)
32. Ji, Z., Liu, K.J.R.: Multi-stage pricing game for collusion-resistant dynamic spectrum allocation. IEEE J. Sel. Areas Commun. **26**(1), 182–191 (2008)
33. Wu, Y., Wang, B., Liu, K.J.R., Clancy, T.C.: A scalable collusion-resistant multi-winner cognitive spectrum auction game. IEEE Trans. Commun. **57**(12), 3805–3816 (2009)
34. Pan, M., Sun, J., Fang, Y.: Purging the back-room dealing: secure spectrum auction leveraging Paillier cryptosystem. IEEE J. Sel. Areas Commun. **29**(4), 866–876 (2011)
35. Chen, Y., Wu, Y., Wang, B., Liu, K.J.R.: Spectrum auction games for multimedia streaming over cognitive radio networks. IEEE Trans. Commun. **58**(8), 2381–2390 (2010)
36. Vickrey, W.: Counterspeculation, auctions, and competitive sealed tenders. J. Financ. **16**(1), 8–37 (1961)
37. Han, Z., Zheng, R., Poor, H.V.: Repeated auctions with Bayesian nonparametric learning for spectrum access in cognitive radio networks. IEEE Trans. Wireless Commun. **10**(3), 890–900 (2011)
38. Zhang, R., Song, L., Han, Z., Jiao, B.: Improve physical layer security in cooperative wireless network using distributed auction games. In: Proceedings of IEEE Conference on Computer Communication Workshops, pp. 18–23 (2011)
39. Stanojev, I., Yener, A.: Improving secrecy rate via spectrum leasing for friendly jamming. IEEE Trans. Wireless Commun. **12**(1), 134–145 (2013)
40. Alavijeh, M.A., Maham, B., Han, Z., Nader-Esfahani, S.: Efficient anti-jamming truthful spectrum auction among secondary users in cognitive radio networks. In: Proceedings of IEEE International Conference on Communication (ICC), pp. 2812–2816 (2013)

41. Resnick, P., Kuwabara, K., Zeckhauser, R., Friedman, E.: Reputation systems. Commun. ACM **43**(12), 45–48 (2000)
42. Gambetta, D.: Can we trust trust. In: Trust: Making and Breaking Cooperative Relations. Basil Blackwell, Oxford, England, pp. 213–237 (2000)
43. Jarvenpaa, S.L., Knoll, K., Leidner, D.E.: Is anybody out there? antecedents of trust in global virtual teams. J. Manag. Inf. Syst. **14**(4), 29–64 (1998)
44. Cahill, V.: Using trust for secure collaboration in uncertain environments. IEEE Pervasive Comput. **2**(3), 52–61 (2003)
45. Theodorakopoulos, G., Baras, J.S.: On trust models and trust evaluation metrics for ad hoc networks. IEEE J. Sel. Areas Commun. **24**(2), 318–328 (2006)
46. Boukerch, A., Xu, L., El-Khatib, K.: Trust-based security for wireless ad hoc and sensor networks. Comput. Commun. **30**(11), 2413–2427 (2007)
47. Xiong, L., Liu, L.: Peertrust: supporting reputation-based trust for peer-to-peer electronic communities. IEEE Trans. Knowl. Data Eng. **16**(7), 843–857 (2004)
48. Chen, K., Peng, Y., Prasad, N., Liang, Y., Sun, S.: Cognitive radio network architecture: part II–trusted network layer structure. In: Proceedings of International Conference on Ubiquitous Information Management and Communication, pp. 120–124 (2008)
49. Clancy, T.C., Goergen, N.: Security in cognitive radio networks: threats and mitigation. In: Proceedings of International Conference on Cognitive Radio Oriented Wireless Networks and Communication, pp. 1–8 (2008)
50. Qin, T., Yu, H., Leung, C., Shen, Z., Miao, C.: Towards a trust aware cognitive radio architecture. ACM SIGMOBILE Mob. Comput. Commun. Rev. **13**(2), 86–95 (2009)
51. Marmol, F.G., G.M. Pérez.: Security threats scenarios in trust and reputation models for distributed systems. Comput. Secur. **28**(7), 545–556 (2009)
52. Gouldner, A.W.: The Norm of Reciprocity: A Preliminary Statement. Am. Sociol. Rev. **25**(2), 161–178 (1960)
53. Seredynski, M., Bouvry, P., Dunlop, D.: Performance evaluation of personal and general data classes for trust management in MANETs. Inf. Media Technol. **6**(3), 936–949 (2011)
54. Dehnie, S., Memon, N.: Modeling misbehavior in cooperative diversity: a dynamic game approach. EURASIP J. Adv. Signal Process. **2009**(3), 1–12 (2009)
55. Sun, Y.L., Yu, W., Han, Z., Liu, K.J.R.: Information theoretic framework of trust modeling and evaluation for ad hoc networks. IEEE J. Sel. Areas Commun. **24**(2), 305–317 (2006)
56. Sun, Y., Han, Z., Liu, K.J.R.: Defense of trust management vulnerabilities in distributed networks. IEEE Commun. Mag. **46**(2), 112–119 (2008)
57. Zhang, N., Yu, W., Fu, X., Das, S.K.: Maintaining defender's reputation in anomaly detection against insider attacks. IEEE Trans. Syst. Man Cybern. B Cybern. **40**(3), 597–611 (2010)
58. Ayday, E., Lee, H., Fekri, F.: Trust management and adversary detection for delay tolerant networks. In: Proceedings of Military Communications Conference, pp. 1788–1793 (2010)
59. Seredynski, M., Bouvry, P.: Direct reciprocity-based cooperation in mobile ad hoc networks. Int. J. Found. Comput. Sci. **23**(02), 501–521 (2012)
60. Asher, D.E., Zaldivar, A., Barton, B., Brewer, A.A., Krichmar, J.L.: Reciprocity and retaliation in social games with adaptive agents. IEEE Trans. Auton. Ment. Dev. **4**(3), 226–238 (2012)
61. Chen, Y., Liu, K.J.R.: Indirect reciprocity game modelling for cooperation stimulation in cognitive networks. IEEE Trans. Commun. **59**(1), 159–168 (2011)
62. Efstathiou, E.C., Frangoudis, P.A., Polyzos, G.C.: Controlled Wi-Fi sharing in cities: a decentralized approach relying on indirect reciprocity. IEEE Trans. Mob. Comput. **9**(8), 1147–1160 (2010)
63. Seredynski, M., Bouvry, P.: Analysing the development of cooperation in MANETs using evolutionary game theory. J. Supercomput. **63**(3), 854–870 (2013)
64. Tang, C., Li, A., Li, X.: When reputation enforces evolutionary cooperation in unreliable MANETs. IEEE Trans. Cybern. **45**(10), 2190–2201 (2014)
65. Gao, Y., Chen, Y., Liu, K.J.R.: Cooperation stimulation for multiuser cooperative communications using indirect reciprocity game. IEEE Trans. Commun. **60**(12), 3650–3661 (2012)

66. Lin, H., Lin, Y., Chang, W.: Reputation auction framework for cooperative communications in green wireless networks. In: Proceedings of IEEE International Symposium Personal Indoor and Mobile Radio Communication, pp. 875–880 (2012)
67. Zhang, B., Chen, Y., Liu, K.J.R.: An indirect reciprocity game theoretic framework for dynamic spectrum access. In: Proceedings of IEEE International Conference Communications (ICC), pp. 1747–1751 (2012)
68. Landa, R., Griffin, D., Clegg, R., Mykoniati, E., Rio, M.: A sybilproof indirect reciprocity mechanism for peer-to-peer networks. In: Proceedings of IEEE International Conference in Computer and Communications (INFOCOM), pp. 343–351 (2009)
69. Xiao, L., Chen, Y., Lin, W.S., Liu, K.J.R.: Indirect reciprocity security game for large-scale wireless networks. IEEE Trans. Inf. Forensics Secur. 7(4), 1368–1380 (2012)

Chapter 6
Active Anti-jamming Solutions in CRNs

6.1 Introduction

With autonomy and control in their radio transmissions, secondary users can perform jamming attacks in cognitive radio networks [1]. Being rational and selfish, an SU can block the ongoing transmissions to maximize its own utility. Extensive works have been done to investigate the impact of jamming on the performance of CRNs. Detection and localization algorithms have been proposed to identify jammers [2–5]. Although most existing works assume that jammers do not concern about their throughput performance, in practice, network access is highly desirable for most users, including many potential jammers. Therefore, the requests for network service are exploited to decrease the attacking rate in CRNs. For instance, rational nodes would hesitate to conduct adversary behavior, if the cost due to the deprivation of network services exceeds the illegal security gain.

Therefore consideration, reciprocity-based anti-jamming communication systems were proposed to stimulate user cooperation and punish jammers [6–14]. For example, the anti-jamming system in [14] punishes insider jammers by stopping their networks services. If the punishment loss exceeds the illegal security gain, rational nodes are stimulated to deviate from jamming for their own interests. The reputation propagation mechanism recognizes and punishes jammers by a much larger node population in the network, compared with direct reciprocity principle. The system based on indirect reciprocity principle provides stronger security protections, especially for large-scale CRNs with node mobility.

As system parameters such as the channel states and the opponent actions are not always known by secondary users, reinforcement learning can be used by secondary users to achieve their optimal power control strategies against jamming. For example, the power allocation strategy based on reinforcement learning proposed in [15] for secondary users resists jamming under unknown channel informations in dynamic CRNs.

© The Author(s) 2015   59
L. Xiao, *Anti-Jamming Transmissions in Cognitive Radio Networks*,
SpringerBriefs in Electrical and Computer Engineering,
DOI 10.1007/978-3-319-24292-7_6

In this chapter, we first present reciprocity mechanisms in CRNs in Sect. 6.2, discuss reputation-based anti-jamming communication systems in Sect. 6.3, and investigate social norms in CRNs in Sect. 6.4. We analyze the evolutionarily stable strategy against jamming in Sect. 6.5 and present anti-jamming transmission strategies based on reinforcement learning in Sect. 6.6. Finally, we conclude in Sect. 6.7.

6.2   Reciprocity in CRNs

With the trust modeling and evaluation method proposed in [16], the trust/reciprocity mechanism stimulates cooperations against jamming in CRNs [6–13]. These works apply direct reciprocity principle to stimulate cooperation among selfish nodes in CRNs. For example, a reputation-based algorithm proposed in [11] forces upcoming attackers to give up attacks and leads to a lower cost in the long run. The punishment based incentive strategy in [12] stimulates cooperations in wireless networks, in which each node monitors others' behavior and makes a similar move to the worst behavior that it has observed. In the CRN based on direct reciprocity principle, each SU chooses its actions according to the interaction history with its opponents, and is more likely to decline the requests from the former attackers. Unfortunately, direct reciprocity principle can effectively defend jamming attacks only under a small SU population without mobility [14].

Indirect reciprocity principle can be applied to address jamming attacks in [7, 14]. The cooperation stimulation scheme based on indirect reciprocity proposed in [7] stimulates cooperation among selfish SUs in large CRNs, which does not rely on the assumption that the interactions between a pair of players are long-lasting. The anti-jamming system based on indirect reciprocity can provide a stronger security protection, compared with the direct reciprocity system, especially for the large-scale CRNs with node mobility [14]. The indirect reciprocity principle is operated by exchanging reputation information among participating entities. In the CRN with indirect reciprocity principle, each secondary user chooses its action according to the reputation of the opponent and is more likely to decline the requests from those who have ever hurt others. As shown in Fig. 6.1, after receiving the request from SU_1 who has ever performed jamming attack to SU_2, SU_3 disobeys the request.

6.3   Anti-jamming System Based on Reputation

The reputation system proposed in [14] addresses jamming in CRNs, in which indirect reciprocity principle is applied to reduce the potential insider jammer population in CRNs. In this system, jammers are not only punished by their direct victims, but also by other SUs in the CRN. Each SU checks the actions of its

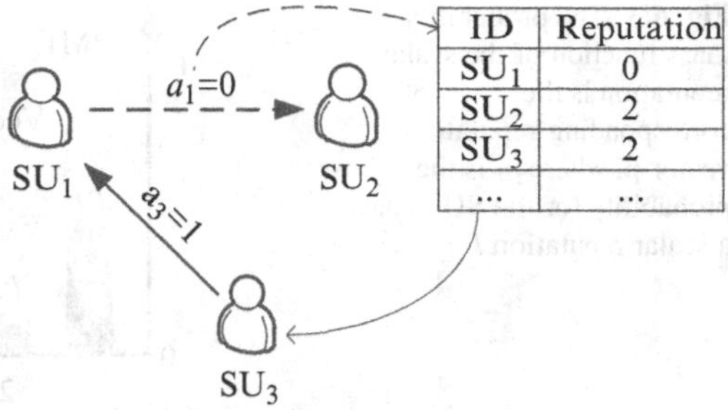

Fig. 6.1 A simple example of the system with indirect reciprocity principle in CRNs, where $a_i \in \{0, 1, 2\}$ denotes that the i-th SU performs jamming attack, declines or follows the transmission request

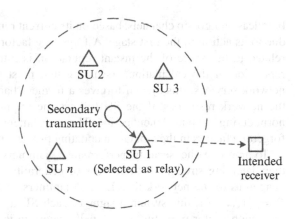

ID	Reputation
SU_1	0
SU_2	2
SU_3	2
...	...

Fig. 6.2 Communication topology of the CRN, including a transmitter, an intended receiver, and n SUs in the communication region (including SU 1 to SU n), with some neighboring SUs selected to relay (SU 1 in this example)

neighbors, updates their reputations, and broadcasts the new reputations to the network via gossip channels.

Each SU is assigned a unique identity that cannot be changed by itself, and knows the identities of its neighbors via local information exchange. Multiple transmissions take place simultaneously in a large-scale network without interfering with each other. The Intended receiver and other observing SUs evaluate the behavior of the neighboring transmitter, update its reputation and propagate the new reputation to the whole network through gossip channels. As shown in Fig. 6.2, each transmission scenario includes a transmitter, an intended receiver, and n neighboring SUs. In each transmission, the transmitter sends a message to the receiver, possibly with the help of some relay SUs.

Each relay SU obtains a reputation vector according to its action history, which determines its opportunity to receive network services. More specifically, the SU with a higher reputation value is more likely to obtain node cooperation in the CRN. On the other hand, the reputation of a relay SU decreases if it attacks the network or disobeys the SU with a good reputation. Meanwhile, the reputation also decreases if the SU helps a "bad" SU. In this way, this system motivates SUs not to perform jamming.

Each transmission consists of two stages: the message transmission stage and the reputation evaluation stage. In the latter stage, an SU's reputation is updated and

Fig. 6.3 The probability
mass function of the scalar
reputation is the
corresponding reputation
vector **p**, where p_l is the
probability for the SU to have
a scalar reputation l

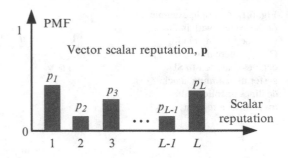

broadcast via gossip channels, based on its current reputation and instant reputation
due to its action in the first stage. A forgetting factor, which can be either fixed or
related to the value of the instant reputation, is introduced to weigh the current
reputation in the calculation. Assuming that most SUs are rational and require
network services, this system forgives a former "bad" SU and allows it to regain
the network resources, if the SU follows the requirement of the network social
norm during the punishment period. The punishment duration is determined by the
forgetting factors in the reputation updating process.

To improve the security performance, jammers are classified, according to
their jamming strength. Without loss of generality, let Level-g jammers be more
dangerous to the network than Level-h jammers, if $g < h$. With a reputation set
$R = \{1, \ldots, L\}$, this system assigns to each SU a scalar reputation $j \in R$ and a
reputation vector $\mathbf{p} = [p_1, p_2, \ldots, p_L]^T$, where p_l is the probability for the SU to
have a scalar reputation l. The scalar reputation j is a realization of an integer random
variable whose probability mass function (PMF) is the corresponding reputation
vector $\mathbf{p}$ as shown in Fig. 6.3. Each SU newly entering the network obtains a high
initial reputation, $\mathbf{p}[0] = [0, \ldots, 0, 1]^T$. In general, an SU whose reputation vector
has a larger mean value is likely to have a higher scalar reputation. Compared
with the scalar reputation that only describes the instant or average value, the
reputation vector includes the likelihood of each reputation, and thus contains more
information on the history of the SU. Therefore, the reputation vector can provide
better performance.

After receiving the request by a transmitter, each SU chooses its action $a_{i,j} \in$
$\{1, \cdots, L\}$ according to its own scalar reputation i and the transmitter's reputation j.
During each transmission, the SUs' reputations are updated based on the same social
norm, which is denoted as $Q = [Q_{i,j}]_{L \times L}$, where $Q_{i,j} \in \{1, \ldots, L\}$ is the instant scalar
reputation assigned to the SU that takes action i towards a transmitter with a scalar
reputation j.

The reputation updating process is illustrated in Fig. 6.4. The instant vector
reputation e_x is the standard basis vector based on the scalar reputation. The new
vector reputation at time $k + 1$ relies on both the instant reputation e_x and the current
reputation $\mathbf{p}[k]$. The latter is weighted by Λ_x, the xth element of the forgetting factor
vector denoted by $\Lambda = [\Lambda_x]_{1 \times L}$. The current action a has less impacts on $\mathbf{p}[k + 1]$,

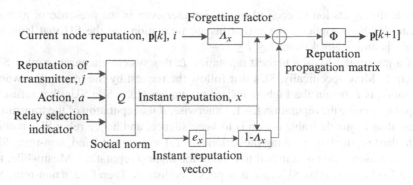

Fig. 6.4 Reputation updating process of the anti-jamming system in [14]

under a larger forgetting factor value Λ_x. The SUs' reputations are then broadcast to the network via gossip channels, with the reputation propagation matrix denoted as $\Phi \triangleq [\Phi_{l,m}]_{L \times L}$, where $\Phi_{l,m}$ is the probability for the reputation m to be taken as l, due to both the behavior detection error and the gossip channel propagation error. The new reputation of the SU propagated via gossip channels is given by

$$\mathbf{p}[k+1] = \Phi(\Lambda_x \mathbf{p}[k] + (1 - \Lambda_x)e_x). \tag{6.1}$$

The security system provides the "good" SUs that have higher reputations with the benefits of receiving better network accesses in the future. On the other hand, jammers are labelled with very low reputations and lose their network services during the punishment time. The punishment duration depends on the forgetting factor in the reputation updating process. If the cost to be punished is larger than the illegal security gain, rational SUs have incentives to choose the desirable actions and abandon adversary behaviors.

6.4 Social Norm in CRNs

As the rules of behavior that are considered to be acceptable in a group or society, social norms represent individuals' basic knowledge of what others do and what others think that they should do. While the economic models of social norms have generated a great deal of interest, it has found widespread use in CRNs [7].

In order to update the SUs' reputations in the reputation-based anti-jamming system, a public social norm denoted as $Q = [Q_{ij}]_{L \times L}$ is designed to guide the SUs' behavior and to suppress jammers in the CRN. In general, SUs can receive high reputations by helping the good SUs or disobeying the transmitters with bad reputations. On the other hand, in order to maintain a healthy network, the system reduces the reputation of each identified jammer even in the presence of a "bad" transmitter. For simplicity, we assign instant reputation i to the SU that launches Level-i jamming attack, with $1 \le i \le L - 2$.

If a transmitter has the highest reputation L, this system encourages other SUs to help it. More specifically, SUs that follow the request by the transmitter whose reputation is L obtain the highest instant reputation (L), while SUs that refuse to cooperate receive the reputation $L - 1$. Otherwise, if the reputation of the transmitter is less than L, the desirable action is to keep silence, and hence relay SUs receive the highest reputation by taking the action $L - 1$. On the other hand, non-relay SUs with the action L do not transmit to receive the highest reputation. Meanwhile, the action $L - 1$ of non-relay SUs results in packet collisions. Therefore, a non-relay SU receives an instant reputation i by taking the action i, with $i \in \{L - 1, L\}$, regardless of the transmitter's reputation.

The relay indicator denoted by ϖ is used to represent the relay selection results that depend on factors such as the network topology, radio channel conditions, and SUs' reputations. Based on the above principles, the social norm can be written as

$$
Q(\varpi) = \varpi \begin{bmatrix} 1 & 1 & \cdots & 1 & 1 \\ 2 & 2 & \cdots & 2 & 2 \\ \cdots & \cdots & \cdots & \cdots & \cdots \\ L{-}2 & L{-}2 & \cdots & L{-}2 & L{-}2 \\ L & L & \cdots & L & L-1 \\ 1 & 2 & \cdots L-1 & L \end{bmatrix} + (1 - \varpi) \begin{bmatrix} 1 & \cdots & 1 & 1 \\ 2 & \cdots & 2 & 2 \\ \cdots & \cdots & \cdots & \cdots \\ L{-}1 & \cdots & L{-}1 & L{-}1 \\ L & \cdots & L & L \end{bmatrix}.
$$

(6.2)

Similarly, the desirable action strategy, denoted as $\hat{\mathbf{a}}$, can be written as

$$
\hat{\mathbf{a}}(\varpi) = \left[\hat{a}_{i,j} \right]_{L \times L} = \varpi \begin{bmatrix} L-1 & \cdots & L-1 & L \\ L-1 & \cdots & L-1 & L \\ \cdots & \cdots & \cdots & \cdots \\ L-1 & \cdots & L-1 & L \end{bmatrix} + (1 - \varpi) \begin{bmatrix} L & L & \cdots & L \\ L & L & \cdots & L \\ \cdots & \cdots & \cdots & \cdots \\ L & L & \cdots & L \end{bmatrix}.
$$

(6.3)

All the SUs in the network know the social norm, the forgetting factor vector, and the reputation updating algorithm. In addition, each SU that updates the reputation for its neighbor is assumed to know the current reputation vector, the relay indicator and the action of the SU under study, as well as the reputation vector of the corresponding transmitter. More specifically, as a heuristic method, the reputation vectors of all the SUs in the network are stored in a central server. Each SU observes the behaviors of its neighboring SUs and updates their reputations according to the reputation updating process, after retrieving their current reputation vectors and the reputation of the transmitter.

6.5 Evolutionarily Stable Strategy Against Jamming

The evolutionarily stable strategy (ESS) can be applied to investigate the stable equilibrium of the proposed anti-jamming system with reputation in [14]. In the reputation-based anti-jamming system, a transmitter selects a subset of n neighboring SUs as its relays. In response, SU i ($1 \leq i \leq n$) chooses an action at time k, denoted as $A_i[k]$, from the action set $\{1, 2, \ldots, L\}$. The action j ($j < L-1$) corresponds to Level-j jamming according to the jamming strength. The action $L-1$ is to disobey the request from the transmitter, while the action L is to follow the request from the transmitter.

In the absence of other SU transmitter, an individual SU taking action j ($1 \leq j \leq L$) can receive an instant payoff, denoted as $C_j(\varpi)$, which is the security gain minus the related cost. Meanwhile, the transmitter with action j receives an instant payoff G_j, which can be the transmission gain or security loss. An SU with a positive payoff gains from the action, while a negative payoff indicates a loss to the SU. Note that the payoff of the player also depends on the relay indicator ϖ. For instance, the action L, i.e., to follow the request, costs more energy to the relay SU, compared with the non-relay SU with $\varpi = 0$. For simplicity of notation, we define $C(\varpi) = [C_i(\varpi)]_{1 \leq i \leq L}$ (or $G = [G_i]_{1 \leq i \leq L}$).

If an SU follows the request by the transmitter, the latter benefits (i.e., $G_L > 0$), while a relay SU has to consume energy to transmit and thus takes a higher cost compared with a non-relay SU, i.e., $C_L(\varpi = 1) < C_L(\varpi = 0) \leq 0$. Note that the anti-jamming system is designed to punish the SUs that attack the network or reject the request by a good SU. Therefore, a rational SU never launches any of those actions unless obtaining a positive instant payoff, i.e., $C_j(\varpi) \geq 0$, for action $j < L$. In this case, the transmitter suffers from the security or throughput loss, implying $G_j \leq 0$, for $j < L$. In addition, the action with a lower label is more dangerous to the network and brings more illegal security advantages to the player itself, and thus $G_j \geq G_i$ and $C_i(\varpi) \leq C_j(\varpi)$, for given $1 \leq i < j \leq L$.

For simplicity, the transmission is assumed to be successful if all the SUs in the area follow the requests. The performance of the transmitter depends on its worst neighbor, or the worst action taken by its n neighbors. For instance, the transmission fails if any neighbor disobeys the request. Another example is that a single jammer can ruin the whole transmission. Therefore, the payoff to the transmitter at time k, denoted as $U_T[k]$, is the minimum instant payoff as follows:

$$U_T[k] = \min_{1 \leq i \leq n} G_{A_i[k]}, \tag{6.4}$$

where $A_i[k]$ is the action taken by SU i at time k. In addition, we denote the payoff of action i to the SU itself as $U_i[k]$ and assume it to be independent of other SUs, i.e.,

$$U_i[k] = C_i(\varpi). \tag{6.5}$$

The intended receiver and the observing SUs monitor the transmission and evaluate the behavior of each SU.

An evolutionarily stable strategy cannot be invaded by any alternative strategy that is initially rare, and natural selection alone is sufficient to prevent alternative strategies from invading. To evaluate the stability property, the Wright-Fisher model [17] is used to study how the action rules spread over the CRN, where the probability for an SU to choose a strategy is proportional to the expected payoff of the population using that strategy. More specifically, the probability for an SU to choose action strategy next time, $y_i[k + 1]$, is given by

$$y_i[k + 1] = \frac{y_i[k] U_i[k]}{\sum_l y_l[k] U_l[k]}. \tag{6.6}$$

The condition that the desired strategy $\hat{\mathbf{a}}$, in (6.3), is viewed as evolutionarily stable, if each SU is motivated to adopt the desired strategy, $\hat{\mathbf{a}} = \mathbf{a}^*$. As indicated by (6.3), SUs select their actions regardless of their own reputations. Therefore, the optimal action in this game can be expressed as

$$\mathbf{a}^*(\varpi) = \begin{bmatrix} a_1^*(\varpi) \\ a_2^*(\varpi) \\ \cdots \\ a_{L-1}^*(\varpi) \\ a_L^*(\varpi) \end{bmatrix} = \varpi \begin{bmatrix} L-1 \\ L-1 \\ \cdots \\ L-1 \\ L \end{bmatrix} + (1 - \varpi) \begin{bmatrix} L \\ L \\ \cdots \\ L \\ L \end{bmatrix} \tag{6.7}$$

where $a_i^*(\varpi)$ is the optimal action of the SU with relay indicator ϖ against a transmitter with a reputation i. The probability for each reputation to be accurately broadcast is denoted by $P_{Di} = \sigma$, where σ is the probability to successfully identify a jammer, and thus the reputation propagation matrix, $\Phi \triangleq [\Phi_{l,m}]_{L \times L}$, can be given by $\Phi_{l,l} = \sigma$ and $\Phi_{l,m} = \frac{1-\sigma}{L-1}, \forall l \neq m$.

We now consider a more interesting case, where each SU is able to perform jamming. In this case, $L = 3$ and the action set is $\{1, 2, 3\}$, whose elements represent jamming attacks, request rejection, and to follow the request by the transmitter, respectively. The corresponding instant payoffs of the SU itself and that of the transmitter are $G = [-8, -1, 10]$ and $C(\varpi = 1) = [5, 1, -0.5]$, respectively. Using (6.2), we obtain

$$Q^{3 \times 3}(\varpi = 1) = \begin{bmatrix} 1 & 1 & 1 \\ 3 & 3 & 2 \\ 1 & 2 & 3 \end{bmatrix} \tag{6.8}$$

and

$$\hat{\mathbf{a}}^3(\varpi = 1) = [2, 2, 3]^T. \tag{6.9}$$

As a benchmark, we also consider an anti-jamming system based on direct reciprocity principle, where each SU chooses its action according to its past interaction with its current opponents. As shown in Fig. 6.5, the system based on direct reciprocity principle fails to work in the networks with $N = 5000$ SUs, and the network corrupts shortly after the start of the process with an eruption of jamming. That is because the long-term punishment cost to a jammer in the direct reciprocity-based approach is small, compared with its illegal security gain, as SUs are unlikely to meet each other again very soon in the large-scale CRN.

Fortunately, the indirect reciprocity-based anti-jamming system can address this problem and efficiently combat attacks in the large-scale CRN. For example, more than 90 % of the population chooses the desirable strategy, shortly after the start of the process in the system, as shown in Fig. 6.5a. As shown in Fig. 6.5c, the system reduces the jammer population from around 5 % to less than 0.05 % shortly after 400 time slots. In addition, as shown in Fig. 6.5, the anti-jamming system that applies $\Lambda = [0.2, 0.4, 0.5]$ has a better security performance than the case with $\Lambda = [0.5, 0.5, 0.5]$. The reason for the significant drop of the jammer population is that the system takes into account the attacking behavior with more weights in the reputation updating, and thus punishes jammers with longer punishment duration.

6.6 Anti-jamming Transmissions Based on Learning

As system parameters such as the channel states and the jamming strategies are not always known in time, reinforcement learning has become an important method for secondary users to choose their transmission strategies to improve their anti-jamming performance. The power allocation strategy proposed in [15] achieves the optimal transmission power and channel selection with unaware parameters such as channel gains based on reinforcement learning.

In CRNs with multiple channels, an SU and a jammer repeatedly choose their power allocation strategies over multiple channels simultaneously without interfering with primary users. Both the SU and jammer are assumed to access only one of M available channels in a time slot and their transmit power is quantized into K levels. The actions of the SU and jammer at the nth time slot are denoted by $\mathbf{x}^n = [P_s, \gamma_s]$ and $\mathbf{y}^n = [P_j, \gamma_j]$, where $P_{s/j} \in \{P_k\}_{1 \leq k \leq K}$ denotes the transmit or jamming power respectively, and $\gamma_{s/j} \in \{1, \cdots, M\}$ is the channel ID selected by the SU (or jammer). The transmission cost per unit power of the SU and jammer are E_s and E_j, respectively.

The channel power gains of the SU (or jammer), denoted by $\mathbf{H}_s = [h_s^{\gamma_s}]_{1 \leq \gamma_s \leq M}$ (or $\mathbf{H}_j = [h_j^{\gamma_j}]_{1 \leq \gamma_j \leq M}$), where $h_s^{\gamma_s}$ (or $h_j^{\gamma_j}$) is the power gain of channel γ_s (or γ_j), is assumed to be constant and known by both nodes. Neither the SU nor jammer is allowed to disrupt the PU's ongoing transmissions. The presence of PUs is indicated by α, where $\alpha = 0$ indicates the presence of PUs on the chosen channel. In each time slot, the SU and the jammer decide their strategies simultaneously based on the network state and the learning history. The SU observes the state, denoted by $\mathbf{s}_s^n = [\alpha^{n-1}, \mathbf{y}^{n-1}]$, and chooses its transmission power and channel ID, denoted

Fig. 6.5 Performance of the reputation-based anti-jamming system with $N = 5000$ SUs, whose transmission probability $\rho = 0.2$, for both the direct reciprocity system and the indirect reciprocity system, with different forgetting factor vectors (Λ) in the reputation updating process, the size of the action set $L = 3$, and the jammer identification rate $\sigma = 0.999$ in [14]. (**a**) Percentage of the population with the desired strategy. (**b**) Total network utility. (**c**) Jamming rate

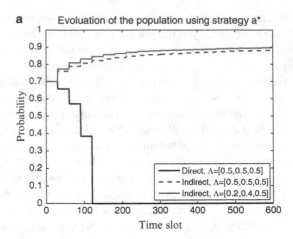

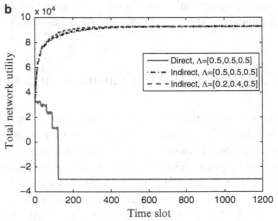

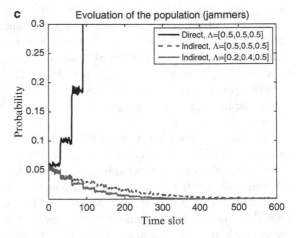

by $\mathbf{x}^n$. The jammer chooses its action $\mathbf{y}^n$ based on the observed state denoted by $\mathbf{s}_j^n = [\alpha^{n-1}, \mathbf{x}^{n-1}]$. Let $I(x)$ is an indicator function.

The immediate utility of the SU at time n, denoted by u_s, is based on the SINR and given by

$$u_s(\mathbf{s}_s^n, \mathbf{x}^n) = \alpha \frac{P_s h_s^{\gamma_s}}{\varepsilon + P_j h_j^{\gamma_s} I(\gamma_s - \gamma_j)} - C_s I(\gamma_s^n - \gamma_s^{n-1}) - E_s P_s, \tag{6.10}$$

where the first term indicates the SINR of the SU, the second term represents the hopping cost of the SU, the last term is the transmission energy consumption of the SU, ε is the constant noise power, C_s is the cost of the channel hopping of the SU, and C_j is the hopping cost of the jammer. Similarly, the immediate utility of the jammer, denoted by u_j, is given by

$$u_j(\mathbf{s}_j^n, \mathbf{y}^n) = -\alpha \frac{P_s h_s^{\gamma_s}}{\varepsilon + P_j h_j^{\gamma_s} I(\gamma_s - \gamma_j)} - C_j I(\gamma_j^n - \gamma_j^{n-1}) - E_j P_j. \tag{6.11}$$

The SU chooses its action $\mathbf{x}^n$ based on state $\mathbf{s}_s^n$ according to its policy denoted by $\pi_s(\mathbf{s}_s^n)$ to receive an immediate utility $u_s(\mathbf{s}_s^n, \mathbf{x}^n)$. Similarly, the jammer chooses its jamming policy $\mathbf{y}^n$ according to $\pi_j(\mathbf{s}_j^n)$, and receives immediate utility $u_j(\mathbf{s}_j^n, \mathbf{y}^n)$. As information such as the state transition probability is not always available to the SU and jammer, Q-learning [18] and WoLF-Q [19] can be used to derive the optimal power allocation strategies.

At the beginning of a time slot, the SU and jammer select their actions $\mathbf{x}^n$ and $\mathbf{y}^n$ simultaneously based on their actions and that of PUs in the last time, respectively. The learning rate of the SU, denoted by μ_s, is firstly updated by

$$\mu_s(\mathbf{s}^n, \mathbf{x}^n) = \frac{1}{1 + N(\mathbf{s}^n, \mathbf{x}^n)}, \tag{6.12}$$

where $N(\mathbf{s}^n, \mathbf{x}^n)$ is the number of the occurrence of the state-action pair $(\mathbf{s}^n, \mathbf{x}^n)$. It is clear that $\mu_s(\mathbf{s}, \mathbf{x})$ decreases over time. The discounting factor, denoted by $\delta \in [0, 1]$, indicates the weight of a future payoff over the current payoff. Let $V_s(\mathbf{s})$ represent the maximum Q value of the SU at state $\mathbf{s}$. The SU's quality function of state-action pair $(\mathbf{s}^n, \mathbf{x}^n)$, denoted by Q_s, is updated by

$$Q_s(\mathbf{s}^n, \mathbf{x}^n) \leftarrow (1 - \mu_s(\mathbf{s}^n, \mathbf{x}^n)) Q_s(\mathbf{s}^n, \mathbf{x}^n) + \mu_s(\mathbf{s}^n, \mathbf{x}^n)(u_s(\mathbf{s}^n, \mathbf{x}^n) + \delta V_s(\mathbf{s}^{n+1}))$$

$$\tag{6.13}$$

$$V_s(\mathbf{s}) \leftarrow \max_{\mathbf{x} \in X} Q_s(\mathbf{s}, \mathbf{x}). \tag{6.14}$$

Algorithm 1 Power Control of SU Based on Q-Learning

1: **Initialize:** $Q_s(\mathbf{s}, \mathbf{x}) = \mathbf{0}$, $V_s(\mathbf{s}) = \mathbf{0}$, $\pi_s(\mathbf{s}) = [P_1, 0, \ldots, 0]'$ and $\mu_s(\mathbf{s}, \mathbf{x}) = \mathbf{1}$, $\forall \mathbf{s}, \mathbf{x}$
2: **for** $n = 1, 2, 3, \ldots$ **do**
3:　　Observe the current state $\mathbf{s}^n$
4:　　Select an action $\mathbf{x}^n$ at random with a small probability η or $\mathbf{x}^n = \pi_s(\mathbf{s}^n)$
5:　　Observe the subsequent state $\mathbf{s}^{n+1}$ and immediate reward u_s
6:　　Update $\mu_s(\mathbf{s}^n, \mathbf{x}^n)$ via (6.12)
7:　　Update $Q_s(\mathbf{s}^n, \mathbf{x}^n)$ via (6.13)
8:　　Update $V_s(\mathbf{s}^n)$ via (6.14)
9:　　Update $\pi_s(\mathbf{s}^n)$ by (6.15)
10: **end for**

Then the SU updates its policy by

$$\pi_s(\mathbf{s}) \leftarrow \arg\max_{\mathbf{x} \in \mathbf{X}} Q_s(\mathbf{s}, \mathbf{x}). \tag{6.15}$$

Details of the SU's channel selection and power control with Q-learning are given in Algorithm 1.

Two learning rates, denoted by δ_{win} and δ_{lose} $(0 <= \delta_{win} < \delta_{lose} <= 1)$, are used by WoLF-Q to compare the current expected value and that of the estimated average policy. If the current expected value is lower (i.e., the player loses), the larger learning rate δ_{lose} is used to learn faster, otherwise δ_{win} is used to learn slower. According to the state $\mathbf{s}^n = [\alpha^{n-1}, \mathbf{y}^{n-1}]$, the number of the occurrence of state $\mathbf{s}^n$, denoted by $Z(\mathbf{s}^n)$, is updated by

$$Z(\mathbf{s}^n) \leftarrow Z(\mathbf{s}^n) + 1. \tag{6.16}$$

Next, the SU estimates the average defense policy, denoted by $\bar{\pi}_{s,w}$, which is given by

$$\bar{\pi}_{s,w}(\mathbf{s}^n, \mathbf{x}) \leftarrow \bar{\pi}_{s,w}(\mathbf{s}^n, \mathbf{x}) + \frac{\pi_{s,w}(\mathbf{s}^n, \mathbf{x}) - \bar{\pi}_{s,w}(\mathbf{s}^n, \mathbf{x})}{Z(\mathbf{s}^n)}. \tag{6.17}$$

The actual defense policy of the SU denoted by $\pi_{s,w}$, is updated by

$$\pi_{s,w}(\mathbf{s}^n, \mathbf{x}) \leftarrow \pi_{s,w}(\mathbf{s}^n, \mathbf{x}) + \triangle_{\mathbf{s},\mathbf{x}}, \tag{6.18}$$

where the weight $\triangle_{\mathbf{s},\mathbf{x}}$ is given by

$$\triangle_{\mathbf{s},\mathbf{x}} = \begin{cases} -\min\{\pi_{s,w}(\mathbf{s}, \mathbf{x}), \frac{\delta_s}{|\mathbf{X}|-1}\}, & \mathbf{x} \neq \arg\max_{\mathbf{x}'} Q_{s,w}(\mathbf{s}, \mathbf{x}') \\ \sum_{\mathbf{x} \neq \mathbf{x}'} \min\{\pi_{s,w}(\mathbf{s}, \mathbf{x}'), \frac{\delta_s}{|\mathbf{X}|-1}\}, & \text{o.w.} \end{cases}, \tag{6.19}$$

where $|\mathbf{X}|$ is the size of $\mathbf{X}$. The learning rate of the SU, denoted as δ_s, is chosen by

$$\delta_s = \begin{cases} \delta_{win}, & \sum_{\mathbf{x}'} \pi_{s,w}^n(\mathbf{s}^n, \mathbf{x}') Q_{s,w}^{n+1}(\mathbf{s}^n, \mathbf{x}') > \sum_{\mathbf{x}'} \bar{\pi}_{s,w}^n(\mathbf{s}^n, \mathbf{x}') Q_{s,w}^{n+1}(\mathbf{s}^n, \mathbf{x}') \\ \delta_{lose}, & \text{o.w.} \end{cases}$$

$$\tag{6.20}$$

Algorithm 2 Power Control of SU Based on WoLF-Q

1: **Initialize:** $Q_s(\mathbf{s}, \mathbf{x}) = 0, V_s(\mathbf{s}) = 0, \pi_{s,w}(\mathbf{s}) = 1/|\mathbf{X}|, \mu_s(\mathbf{s}, \mathbf{x}) = 1$ and $Z(\mathbf{s}) = 0, \forall \mathbf{s}, \mathbf{x}$.
2: **for** $n = 1, 2, 3, \ldots$ **do**
3:     Observe the current system state $\mathbf{s}^n$
4:     Select an action $\mathbf{x}^n$ at random with a probability η or the probability policy $\pi_{s,w}(\mathbf{s}^n, \mathbf{x}^n)$
5:     Observe the subsequent state $\mathbf{s}^{n+1}$ and immediate reward u_s
6:     Update $\mu_s(\mathbf{s}^n, \mathbf{x}^n)$ via (6.12)
7:     Update $Q_s(\mathbf{s}^n, \mathbf{x}^n)$ via (6.13)
8:     Update $V_s(\mathbf{s}^n)$ via (6.14)
9:     Update $Z(\mathbf{s}^n)$ via (6.16)
10:    Update $\bar{\pi}_{s,w}$ via (6.17)
11:    Update $\pi_{s,w}$ via (6.18)
12: **end for**

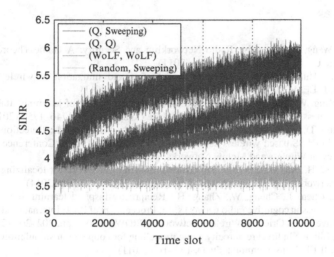

Fig. 6.6 SINR of an SU with anti-jamming power control strategies based on Q-learning and WoLF-Q with $\alpha = 1, \delta = 0.7, \delta_{win} = 0.05, \delta_{lose} = 0.1, M = 16, \epsilon = 1, C_s = C_j = 0.01, E_s = E_j = 0.01$ and $P_i \in \{0, 5, 10\}$ in [15]

Details of the SU's channel hopping and power control algorithm with WoLF-Q are presented in Algorithm 2. To investigate the rarely accessed state-action pairs, the SU applies η-greedy algorithm, i.e., one of other actions is chosen with a small probability η in the update.

Figure 6.6 indicates the performance of the power control strategy with Q-learning against a sweeping jammer with jamming power $P_j = 10$, transmission cost per unit power $E_j = 0.01$ and hopping cost $C_j = 0.01$. The SINR increases with time and converges to the optimal value at about 5.7, due to the SU's capability to fast learn the jammer. If both the SU and jammer use WoLF-Q, the SINR of the SU increases soon after the start of the game and eventually converges to about 4.5.

6.7 Summary

In this chapter, we have investigated active anti-jamming techniques for CRNs by exploiting the requirement of network access by secondary users. By punishing jammers with different strengths, a secure cognitive radio network can suppress the jamming motivations of CR nodes to reduce the jammer population. If the system parameters such as channel states and opponent actions are unknown, the power allocation strategies based on reinforcement learning can improve the SINRs and utilities of secondary users.

References

1. Liu, K., Wang, B.: Cognitive Radio Networking and Security: A Game Theoretical View. Cambridge University Press, Cambridge (2011)
2. Chiang, J., Hu, Y.: Cross-layer jamming detection and mitigation in wireless broadcast networks. IEEE/ACM Trans. Networking 19(1), 286–298 (2011)
3. Lu, Z., Wang, W., Wang, C.: Modeling, evaluation and detection of jamming attacks in time-critical wireless applications. IEEE Trans. Mobile Comput. 13(8), 1746–1759 (2014)
4. Giustiniano, D., Lenders, V., Schmitt, J., Spuhler, M., Wilhelm, M.: Detection of reactive jamming in DSSS-based wireless networks. In: Proceedings of ACM Conference on Security and Privacy in Wireless and Mobile Networks, pp. 43–48 (2013)
5. Liu, Z., Liu, H., Xu, W., Chen, Y.: An error-minimizing framework for localizing jammers in wireless networks. IEEE Trans. Parallel Distrib. Syst. 25(2), 508–517 (2014)
6. Chen, X., Chen, T., Cheng, W., Zhang, H.: Reciprocity inspired learning for opportunistic spectrum access in cognitive radio networks. In: Proceedings of IEEE International Conference on Cognitive Radio Oriented Wireless Networks (CROWNCOM), pp. 202–207 (2013)
7. Chen, Y., Liu, K.: Indirect reciprocity game modelling for cooperation stimulation in cognitive networks. IEEE Trans. Commun. 59(1), 159–168 (2011)
8. Sun, Y., Han, Z., Liu, K.: Defense of trust management vulnerabilities in distributed networks. IEEE Commun. Mag. 46(2), 112–119 (2008)
9. Sun, Y., Han, Z., Yu, W., Liu, K.: A trust evaluation framework in distributed networks: vulnerability analysis and defense against attacks. In: Proceedings of IEEE INFOCOM, pp. 1–13 (2006)
10. Yu, W., Liu, K.: Game theoretic analysis of cooperation stimulation and security in autonomous mobile ad hoc networks. IEEE Trans. Mobile Comput. 6(5), 459–473 (2007)
11. Zhang, N., Yu, W., Fu, X., Das, S.: Maintaining defender's reputation in anomaly detection against insider attacks. IEEE Trans. Syst. Man Cybern. B Cybern. 40(3), 597–611 (2010)
12. Niu, B., Zhao, H., Jiang, H.: A cooperation stimulation strategy in wireless multicast networks. IEEE Trans. Signal Process. 59(5), 2355–2369 (2011)
13. Ayday, E., Lee, H., Fekri, F.: Trust management and adversary detection for delay tolerant networks. In: Proceedings of IEEE Military Communication Conference, pp. 1788–1793 (2010)
14. Xiao, L., Chen, Y., Lin, W., Liu, K.: Indirect reciprocity security game for large-scale wireless networks. IEEE Trans. Inf. Forensics Secur. 7(4), 1368–1380 (2012)
15. Chen, T., Liu, J., Xiao, L., Huang, L.: Anti-jamming transmissions with learning in heterogenous cognitive radio networks. In: Proceedings of IEEE Wireless Communication and Networking Conference, pp. 235–240 (2015)

16. Sun, Y., Yu, W., Han, Z., Liu, K.: Information theoretic framework of trust modelling and evaluation for ad hoc networks. IEEE J. Sel. Areas Commun. **24**(2), 305–317 (2006)
17. Fisher, R.: The Genetical Theory of Natural Selection. Cambridge University Press, Cambridge (1930)
18. Watkins, C., Dayan, P.: Q-learning. Mach. Learn. **8**(3–4), 279–292 (1992)
19. Bowling, M., Veloso, M.: Multiagent learning using a variable learning rate. Artif. Intell. **136**(2), 215–250 (2002)

Chapter 7
Conclusion and Future Work

7.1 Summary of Jamming in CRNs

We have investigated jamming attacks in cognitive radio networks based on game theory and learning techniques. We reviewed traditional anti-jamming techniques in Chap. 2 and discussed in Chap. 3 how to improve the anti-jamming performance of spread spectrum-based systems. More specifically, uncoordinated spread spectrum techniques resist smart jammers without requiring any pre-shared PHY-layer keys and node cooperation is applied to improve both jamming resistance and communication efficiency. The spectrum efficiency of the anti-jamming transmissions was analyzed and the implementation issues were discussed.

In Chap. 4, we applied game theory to investigate the interactions between the jammer and the transmitter in cognitive radio networks. Both the NE and Stackelberg Equilibrium were presented for one-shot games and then repeated anti-jamming games were discussed. We also studied the impact of node cooperation and inaccurate information on the performance of the anti-jamming transmission games.

We reviewed game theoretic cooperation stimulation methods in Chap. 5, and investigated active anti-jamming solutions in CRNs based on reciprocity principles in Chap. 6. We discussed evolutionarily stable anti-jamming strategies and then presented the anti-jamming strategies based on reinforcement learning for dynamic networks with nodes being unaware of the system parameters.

© The Author(s) 2015
L. Xiao, *Anti-Jamming Transmissions in Cognitive Radio Networks*,
SpringerBriefs in Electrical and Computer Engineering,
DOI 10.1007/978-3-319-24292-7_7

7.2 Future Work

The anti-jamming communications in cognitive radio networks described here suggest a number of research topics. To cite just a few:

1. Most existing game theoretic study on anti-jamming communications in CRNs is based on simplified network models and channel models. For instance, the channel gains between the legal transmitter and receiver are usually assumed to be known by the jammer, which does not always hold in practice. An important future topic is to formulate a jamming game that consists of more detailed and accurate assumptions.
2. Most existing learning-based anti-jamming techniques depend on trial-and-error manners. However, the convergence speed of the anti-jamming transmissions can significantly increase if incorporating known channel models and parameters.
3. In game theoretic study on jamming attacks in cognitive radio networks, the performance is validated via numerical analysis and simulations. In the future, the proposed algorithms have to be implemented in actual cognitive radio systems, such as USRP to derive more accurate results.

Printed in the United States
By Bookmasters